Mathematics and Humor

John Allen
Paulos

Mathematics and Humor

The University of Chicago Press
Chicago and London

JOHN ALLEN PAULOS received his Ph.D. in mathematics from the University of Wisconsin in 1974 and is now associate professor of mathematics at Temple University.

The University of Chicago Press, Chicago 60637
The University of Chicago Press, Ltd., London

Printed in the United States of America
84 83 82 81 5432

Library of Congress Cataloging in Publication Data

Paulos, John Allen.
Mathematics and humor.

Bibliography: p.
Includes index.
1. Wit and humor—Philosophy. I. Title.
PN6149.P5P3 801′.957 80–12742
ISBN 0–226–65024–3

To the
memory of
Abraham
Schwimmer

Contents

1

Mathematics and Humor

The Talmud says, "Begin a lesson with a humorous illustration." This is especially apt here, since discussions of humor are often ponderous and grim, and since this one, though of a quite different sort, is not altogether an exception. A friend of mine—a mathematician, incidentally—recently completed a speed-reading course, and he noted this in a letter to his mother. His mother responded with a long, chatty letter in the middle of which she wrote, "Now that you've taken that speed-reading course, you've probably already finished reading this letter." What this illustrates about my friend's mother may be clearer than what it illustrates about mathematics and humor, but then I have a whole book to explain the latter.

A good way to begin that explanation is to provide a very brief sampling of what philosophers, psychologists, writers, and critics have said in attempting to come to an understanding of humor—hence the following chronology.

Classical writers on humor and laughter considered them base and ignoble. Aristotle, who devoted many pages of his *Poetics* to tragedy, had relatively little to say about comedy (at least relatively little that has survived). He wrote, "Comedy, as we have said, is a representation of inferior people, not indeed in the full sense of the word 'bad,' but the laughable is a species of the base or ugly. It consists in some blunder or ugliness that does not cause pain or disaster, an obvious example being the comic mask which is ugly and distorted but not painful." Plato in *Philebus* wrote that, in laughter, pain (often in the form of envy) and pleasure are mixed. Similarly, Cicero stated that "the province of the ridiculous . . . lies in a certain baseness and deformity."

The ancient conception of humor, of course, was narrower than ours, being limited largely to what we would call farce, burlesque, and slapstick and excluding "higher" forms that might have raised the classical estimate. In the plays of Aristophanes and other Greek comic playwrights, for example, clowns would wander around the stage making obscene gestures.

Up until the seventeenth century, writers on humor were content to more or less repeat the classical formulations, despite the wealth of humor (broadly conceived so as to include Shakespeare, Rabelais, Chaucer, etc.) written during the intervening millennium and a half. The English philosopher Thomas Hobbes introduced in his *Leviathan* (1651) a theory of laughter, usually referred to as the superiority or disparagement theory, that in some restricted form or other has been adopted by many subsequent theorists. He wrote, "Sudden glory is the passion which maketh those grimaces called laughter; and is caused either by some sudden act of their own that pleaseth them; or by apprehension of some deformed thing in another, by comparison whereof they suddenly applaud themselves." Though this feeling of self-satisfied superiority and gloating is a factor in many kinds of humor, it plays a dominant role, I think, only in sick jokes, certain kinds of ethnic jokes, and so forth. It is the primitive base out of which, or beyond which, more "refined" types of jokes and humor have developed. Anybody who denies its existence is sick and should be severely beaten.

The next step in this abbreviated survey is the eighteenth-century Scottish poet and philosopher James Beattie, who made a major study of humor and laughter (1776) in which he wrote: "Laughter arises from the view of two or more inconsistent, unsuitable, or incongruous parts or circumstances, considered as united in complex object or assemblage, or as acquiring a sort of mutual relation from the peculiar manner in which the mind takes notice of them." Besides being the first person to clearly enunciate the so-called incongruity theory of humor, (John Locke wrote less explicitly of similar ideas a little earlier), Beattie was also one of the first to note that laughter and mild fear, as in nervous giggling, are often associated.

This idea that incongruity (oddness, inappropriateness) is at the base of humor was developed in the late eighteenth and early nineteenth centuries by the critic Hazlitt and the philosophers Schopenhauer and Kant. Hazlitt wrote (1819), "The essence of the laughable is the incongruous, the disconnecting of one idea from another, or the jostling of one feeling against another." Kant emphasized the element of surprise, the unexpectedness of the incongruity. "Laughter," he said in a famous formulation (1790), "is an affectation arising from the sudden transformation of a strained expectation into nothing." Finally, Schopenhauer wrote (1818) that humor "often occurs in this way: two or more real objects are thought through one concept; it then becomes strikingly apparent from the entire difference of the objects in other respects, that the concept was only applicable to them from a one-sided point of view." That the words "odd" and "funny" have come to be used interchangeably in many contexts testifies to the naturalness of incongruity accounts of humor. The incongruity of linking names like Kant and Schopenhauer (the "gloomy" pessimist) with notions like humor and laughter may strike the reader as itself a little funny.

A new idea was added to the literature on humor later in the century by Herbert Spencer, who reasoned that the laughter that often (but not always) accompanies amusement is due to an overflow of surplus energy through the facial muscles and respiratory system. It results when the serious expectations of the person laughing are not met and his attention is diverted to something frivolous—or, to quote Spencer, "when consciousness is unawares transferred from great things to small." The redundant psychic "energy" has nowhere to go and so comes out as laughter. Darwin also commented on the physiological basis of laughter, and the idea of unnecessarily generated energy being drained off in laughter influenced many later theorists, in particular Freud.

The incongruity theory and the disparagement theory were put forth ("rediscovered" might be a better word) by a number of theorists in the late nineteenth and early twentieth centuries. George Meredith, a nineteenth-century literary critic, emphasized a different aspect of humor. In *An Essay on Comedy* (1918 edition) he wrote that the "Comic Spirit" is a sort of social corrective and springs to action whenever men "wax out of proportion, overblown, affected, pretentious, bombastical, hypocritical, pedantic; whenever it sees them self-deceived or hood-winked, given to run riot in idolatries . . . planning shortsightedly, plotting dementedly." Many other writers have since pointed out this regulatory aspect of humor. Meredith also wrote that humor, societal health, and the social equality of women and men were all closely related. Actually, this last idea almost follows from his conception of the "Comic Spirit," since "pretentious, bombastical" men would be more easily deflatable in relatively nonsexist societies. (Many studies have shown what common sense suggests—that one's attitudes toward the opposite sex (or opposite sexes) can easily be determined by the types of jokes one finds funny.)

The French poet Baudelaire expressed eloquently the notion that laughter is induced by the realization that we are both physical and spiritual creatures, that we have a sense of both the ridiculous and the sublime. Laughter, he wrote (1868), "is the consequence in man of the idea of his own superiority. And since laughter is essentially human, it is, in fact, essentially contradictory; that is to say that it is at once a token of an infinite grandeur and an infinite misery—the latter in relation to the absolute Being of whom man has an inkling, the former in relation to the beasts. It is from the perpetual collision of these two infinities that laughter is struck. The comic and the capacity for laughter are situated in the laugher and by no means in the object of the laughter." This last sentence begins to get at the complexity of humor —the human laugher and his intentions, values, and so forth.

Coming finally to the twentieth century, the French writer Bergson (1911) attributed laughter to the "mechanical encrusted on something living." By this rather celebrated phrase he meant that when man becomes rigid, machinelike, and repetitive he becomes laughable, since the essence of humanity is its flexibility and spirit. The following quotations to that effect from Bergson are very similar to Baudelaire's. "Any incident is comic that calls our attention to the physical in a person, when it is the moral side that is concerned" (imagine what examples you will). "We laugh everytime a person gives the impression of being a thing." Bergson also put forward the idea that "a momentary anaesthesia of the heart," a certain disinterestedness or lack of sympathy is necessary for the appreciation of humor whose "appeal is to intelligence, pure and simple." Consider animated cartoons where the terrible "suffering" of animals—falling off cliffs, having things explode in their faces—is funny. His theory, unlike most, seems to have been strongly influenced by his reading of humor and comedy. Molière, whose humor is largely due to characters with humorous fixations, quirks, and rigidities, was a particular such influence.

At the risk of being a bit rigid and fixated myself, I will continue this chronological listing of contributions to a theory of humor with a very well-known one. Freud's theory of wit and humor is treated in his *Jokes and Their Relation to the Unconscious* (1905) and is an integral part of his theory of psychoanalysis. Very briefly and somewhat simplistically summarized, Freud's theory maintains that jokes or witticisms enable a person to vent his aggressive or sexual feelings and anxieties in a disguised, subdued, even playful manner. "Tendency wit" must cloak its aggressive or sexual content so as to disarm one's conscience (superego) and allow release of repressed psychic energy. Ambiguity, double meanings, and puns are on this view merely the devices necessary to placate the censoring superego. The repressed energy released in this way takes the form of laughter. Freud also acknowledged the existence of what he called "harmless wit," a joke not carrying any emotional charge. An example of a Freudian joke, albeit a little crude, is the following. Man: "What part of my anatomy is so long and hard and sticks so far out of my pajamas that my hat can be supported on it?" Woman is politely evasive. Man: "My head."

A lesser known but, I think, more insightful humor theorist is the writer Max Eastman, who is one of several people who have emphasized the continuity of humor with play and stressed the importance of a playful, disengaged attitude to the appreciation of humor. He writes (1936): "An atom of humor is an unpleasantness or a frustration taken playfully. A witty joke is made by combining this unpleasantness or frustration with some idea or attitude of feeling in which one can find momentary satisfaction."

Eastman also developed the "derailment" theory of humor. Humorous comments, happenings, and so forth, are incongruous not per se, but only given the context in which they occur. The normal flow of things is "derailed" by them. Most everyday humor is of this context-sensitive kind and thus is as varied as the contexts and situations in which it appears. Characterizing such humor, requiring as it does reference to people's purposes and values as well as to their situations and roles, is usually difficult even in particular cases and is probably impossible in general. Its aliveness contrasts with the staleness of most "canned jokes."

In a somewhat similar vein D. H. Monro, in his *Argument of Laughter* (1951), states that delight in what is new and fresh and a desire to escape from boredom and monotony are important aspects of what is meant by a sense of humor. Humor in which this freshness, novelty, and playfulness are important factors is generally more sophisticated than, say, humor deriving from ethnic or "dirty" jokes, although not necessarily funnier; compare Lewis Carroll with the Three Stooges.

Recently Arthur Koestler in *The Act of Creation* (1964) has emphasized the continuity of creative insights in humor with creative insights in science and poetry. "The logical pattern of the creative process is the same in all three cases: it consists in the discovery of hidden similarities. But the emotional climate is different . . . the comic simile has a touch of aggressiveness; the scientists' reasoning by analogy is emotionally detached, i.e., neutral; the poetic image is sympathetic or admiring, inspired by a positive kind of emotion." Koestler's theory of humor is an incongruity theory that also accounts for the psychological aspects of humor. He maintains that humor results from the "bisociation" of two incompatible frames of reference and that laughter is due to the discharge of emotional energy that, "owing to its greater mass momentum, is unable to follow the sudden switch of ideas to a different type of logic or a new rule of the game; less nimble than thought it tends to persist . . . and finds its outlet in laughter" (cf. Spencer).

As I mentioned before, this is just a sample of what philosophers, psychologists, and critics have written concerning humor and laughter. More recently, much experimental work has been done by social scientists trying to confirm, refine, unify, and extend some of these ideas. Philosophers also, though not directly concerned with humor, have in recent decades clarified concepts relating to human actions and language that are of value in understanding humor. Some of this psychological and philosophical work is fascinating and will be referred to later, but my emphasis here will be on the logic and mathematics of humor, about which almost nothing has been written. By this I do not mean, of course, an analysis differentiating comedy, farce, satire, and so forth. Neither do I mean—although this is closer, and will be returned to later—an analysis of various comic devices, personages, or myths used from Aristophanes through Shakespeare to today's situation comedies. What I do propose to do is to explore the operations and structures common to humor and the formal sciences (logic, mathematics, and linguistics) and to show that various notions from these sciences provide formal analogues for various sorts of jokes and joke patterns. Moreover, in chapter 5 I will develop a mathematical model of jokes (to a certain extent of humor in general) using notions from the mathematical theory of "catastrophes." From time to time relevant philosophical and psychological matters will also be discussed to provide a broader context for the technical ideas developed.

Let me stress that reducing humor to formulas and equations is not my goal. Humor, though often utilizing various formal devices, depends ultimately on meaning that cannot be reduced in this way.

I will not assume that readers have any special background in mathematics, and I will therefore spend considerable time developing the required mathematical ideas. In fact, an auxiliary purpose will be to develop these necessary ideas in a manner more pleasant than the one in which they are usually encountered. Part of the book can in a sense be considered a detailed case study of Koestler's principle that creative insights in all fields (mathematics and humor in this case) share the same logical pattern.

Some loose definition of humor will be helpful before I go on. If one rereads the excerpts I have quoted or looks at other writings on the subject, one finds that two major strands run through most of them—incongruity and the psychological aspect of humor.

Most of the theorists I have cited (as well as those not quoted here) agree, once allowance is made for different ways of putting things and different emphases, that a necessary ingredient of humor is that two (or more) incongruous ways of viewing something (a person, a sentence, a situation) be juxtaposed. In other words, for something to be funny, some unusual, inappropriate, or odd aspects of it must be perceived together and compared. We have seen that different writers have emphasized different oppositions: expectation versus surprise, the mechanical versus the spiritual, superiority versus incompetence, balance versus exaggeration, and propriety versus vulgarity. I will henceforth use the word "incongruity" in an extended sense comprising all the above oppositions.

Incongruity by itself is not, however, a sufficient condition for humor for three reasons: (1) it may not be noticed; (2) it may not have a point or be reasonably resolvable; and (3) the "emotional climate" may not be right. Thus, for example, a play on words may contain a very subtle and therefore unnoticed incongruity, or the absurdity of a given situation may not be realized for one reason or another. Regarding item (2), snow in May is incongruous yet has no point (meaning, gist, nub). Neither does the juxtaposition of an apple and a screwdriver. Determining whether some combination is incongruous and, if it is, whether it has a point, is in general easy to do but quite difficult (perhaps impossible) to describe how to do. I will return to this problem later.

The proper psychological or emotional climate is another essential ingredient of humor. This is also difficult to characterize, but it is clear even from the excerpts quoted that a subdued sort of aggression or self-satisfaction is often present. The aggressive tone may be very slight (sometimes even completely absent). Similarly, the self-satisfaction may result not from "sudden glory" at one's superiority or at another's infirmity but from the overcoming of a mild fear or anxiety or the resolving of an ambiguity (as in figuring out a riddle or pun). A playful, unimpassioned frame of mind also seems to be required. Laughter, if there is any, can be considered to result from the energy dissipated by the punch line of the story.

Together then, two ingredients—a perceived incongruity with a point and an appropriate emotional climate—seem to be both necessary and sufficient for humor. This definition is admittedly rather loose, but it is tight enough for my purposes now. (I will get back to it near the end of the book.) I will not say too much about what constitutes an appropriate emotional climate,[1] but I will, as I mentioned, try to show how notions from mathematics, logic, and linguistics provide formal analogues for certain types of jokes (perceived incongruities with a point) and joke patterns as well as exploring operations and structures common to humor and mathematics.

Before I begin my development of these patterns, operations, and structures in chapter 2, I would like to discuss some similarities of a general sort between mathematics and humor. I became interested in the relation between mathematics and humor when I noticed that mathematicians often had a distinctive sense of humor. What made it distinctive was unclear at first, and so I searched for similarities between mathematical thought and humor.

Both mathematics and humor are forms of intellectual play, the emphasis in mathematics being more on the intellectual, in humor more on the play. To a great degree, combinations of ideas and forms[2] are put together and taken

1. There is an ample body of literature, much of it quite trivial, on this psychological aspect of humor.

2. Though formulas, equations, and computation are essential to mathematics, they are not nearly as important as the mathematical ideas and structures they are intended to partially capture.

apart just for the fun of it. Both activities are undertaken for their own sake. Ingenuity and cleverness are hallmarks of both. Of course I am speaking here of pure mathematics—the art and science of abstract pattern and structure—and not of computational mathematics, which is more a collection of techniques. I am also referring to "pure humor." The analogue to computational mathematics might be, I suppose, manipulative uses of humor in public relations, advertising, and promotion.

Logic, pattern, rules, structure—all these are essential to both mathematics and humor, although of course the emphasis is different in the two. In humor the logic is often inverted, patterns are distorted, rules are misunderstood, and structures are confused. Yet these transformations are not random and must still make sense on some level. Understanding the "correct" logic, pattern, rule, or structure is essential to understanding what is incongruous in a given story—to "getting the joke."

In addition, both mathematics and humor are economical and explicit. Thus the beauty of a mathematical proof depends to a certain extent on its elegance and brevity. A clumsy proof introduces extraneous considerations; it is long-winded or circuitous. Similarly, a joke loses its humor if it is awkwardly told, is explained in redundant detail, or depends on strained analogies.

The logical technique of reductio ad absurdum is important enough to both humor and mathematics to warrant its own paragraph or two. It is a favorite gambit in mathematical proofs and, simply stated, comes to the following. To prove statement S, it is enough to assume the negation of S (not S) and from the negation derive a contradiction. It is probably the prevalence of this technique and of logic in general in mathematics that partially accounts for the propensity of mathematicians to develop all the absurd consequences of any statement offered them. Being in the habit of taking statements literally also contributes, since the literal and figurative interpretations are usually incongruous.

Humor can easily be contrived in this manner. An odd premise is accepted, and the joke or story develops the premise to the point of absurdity. Or a reasonable but figuratively expressed statement is interpreted literally and developed accordingly. For example, innumerable humorous stories have a beginning paragraph whose gist is "What would happen if . . . ," where ". . ." is the premise whose absurd consequences are developed in the story. The emphasis again is different in humor than in mathematics. In humor, reducing the premise to absurdity is usually done more for the sake of the absurdity than to refute the original premise. Often though, as in satire, both motives are present.

Some elementary examples of mathematical proofs are needed here to illustrate the aspects of mathematics mentioned above and to serve as illustrations for further discussion.

One of the most important theorems in Euclidean geometry is the Pythagorean theorem, which states that the square on the hypotenuse of a right triangle is equal to the sum of the squares of its legs (fig. 1).

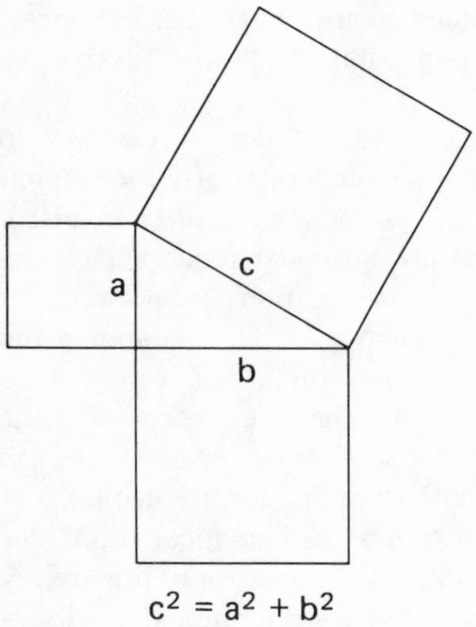

Fig. 1

Among the many proofs of this theorem, the following "pictorial" proof is especially elegant. Consider a right triangle (as above, with sides a, b, and hypotenuse c) and a square whose sides are of length $a + b$. Arrange four of the given right triangles in this square in the two ways shown in figure 2. The area remaining in the square after subtracting the area of the four triangles is c^2 in one case and $a^2 + b^2$ in the other.

I spoke above about economy, elegance, and intellectual play. These qualities should be a little clearer in the light of this beautiful proof. The figure itself is a sort of "punch line" to a very rarefied "joke."

The following two combinatorial results are also illustrative. Consider the problem posed to probably the greatest mathematician who ever lived, Karl Friedrich Gauss, when he was in primary school. (It is strange that no educated person will admit to being completely ignorant of Shakespeare—probably the greatest writer who ever lived, assuming the title has some meaning—yet very few "educated" people are reluctant to admit their ignorance of Gauss, Euler, Poincaré, etc.) The teacher, to quiet the class for a while, asked them to find the sum of the first hundred integers. Gauss almost immediately replied with 5,050. What he did is clear from figure 3.

There are fifty pairs of numbers, each pair equal to 101, and 50 times 101 is 5,050. The same idea works in general, and the following formula is thereby obtained: $1 + 2 + 3 + \ldots n = \frac{n}{2}(n + 1)$. Again, a clever insight suddenly allows us to grasp the solution at a glance.

Let us examine now the problem of finding a way to cover with thirty-one dominoes a checkerboard with two diagonally opposite corners removed (that is, a checkerboard having sixty-two rather than sixty-four squares). Try it before reading on (fig. 4). One way to proceed is to start covering the board with dominoes and see what can be done. Another approach is to make an easy but very penetrating observation: every domino covers one black square and one white square. Thus, since the two missing squares are both white, there is no way to cover the remaining sixty-two squares with thirty-one dominoes!

Finally, let me prove that there are infinitely many prime numbers. The proof, due to Euclid, is a beautiful example of a proof by reductio ad absurdum. Recall that a prime number is any number whose only divisors are itself and 1. Thus 2, 3, 5, 7, 11, 13, 17, 19, 23, and 29 are the first ten prime numbers. If one continues to list the prime numbers, one notices that they become more and more sparsely placed. We want to prove that nevertheless there are infinitely many of them. We thus assume that there are only finitely many of them and try to derive a contradiction from this assumption. Thus, we list the prime numbers 2, 3, 5, 7, . . . p; p we will

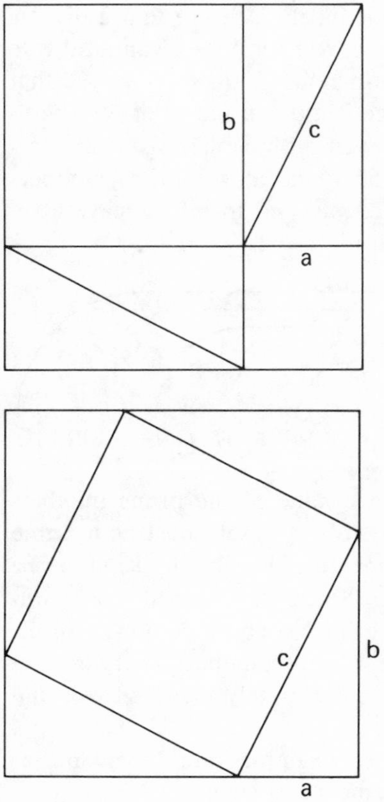

Fig. 2

take to be the largest prime number. (Since we are assuming there are only finitely many prime numbers, there must be a largest.) Now we form a new number N by multiplying all the primes in the above list together. Thus $N = 2 \cdot 3 \cdot 5 \ldots p$.

Now let us consider the number $N + 1$ and see whether 2 divides it evenly (with no remainder). We see that 2 divides N evenly, since it is a factor of N. Therefore 2 cannot divide $N + 1$ evenly, since there is a remainder of 1. We see that 3 divides N evenly also, since it too is a factor of N. Therefore 3 cannot divide $N + 1$ evenly, since there is again a remainder of 1. Similarly for 5, 7, and all the prime numbers up to p. They all divide N evenly and therefore leave a remainder of 1 when divided into $N + 1$.

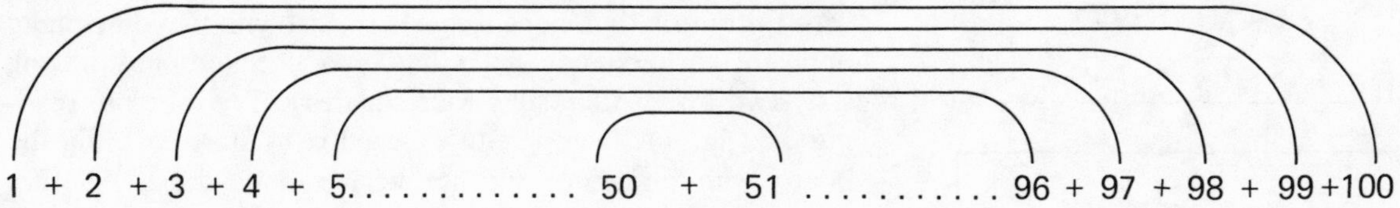

1 + 2 + 3 + 4 + 5. . . . . 50 + 51 . . . . 96 + 97 + 98 + 99+100

Fig. 3

What does this mean? Since none of the prime numbers 2, 3, 5, . . . p divides $N + 1$, $N + 1$ itself must be a prime number larger than p, or it must be divisible by some prime number that is larger than p. Since we assumed that p was the largest prime number, we have a contradiction: a prime number larger than the largest prime number. Therefore our original assumption that there are finitely many prime numbers must be false.

Let me reiterate that these examples are by no means meant to be funny; they are meant to show that some qualities inherent in a good mathematical proof are similar to qualities inherent in good humor: cleverness and economy, playfulness, combinatorial ingenuity, and logic (particularly reductio ad absurdum).

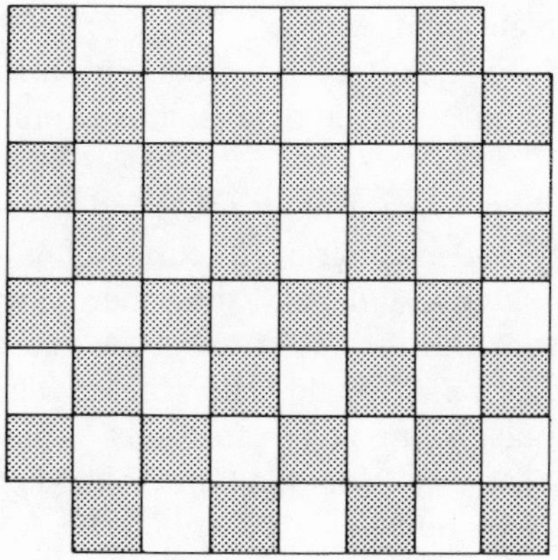

Fig. 4

The assertive tone of much humor seems to be largely absent in mathematics, where a more neutral or positive attitude is more common. Nevertheless, one should remember that one of the motivations of the early Greeks in inventing and refining correct canons of logical argument was the competitive desire to defeat an opponent in debate. This competitiveness is without doubt still very much a factor in the psychology of most mathematicians. In fact, it sometimes happens that in a mathematical seminar everyone in the room is trying to prove the same thing: "I am the best mathematician in the room."

Riddles, trick problems, paradoxes, and "brain teasers"[3] seem to be a bridge between humor and mathematics—more intellectual than most jokes, lighter than most mathematics. Consider as an example the following well-known problem. Two locomotives begin 300 miles apart, heading toward each other on the same track. The first locomotive travels at 100 miles an hour and the second travels at 50 miles an hour. As the locomotives depart, a bird flying 200 miles an hour leaves the first locomotive and heads for the other one. Upon reaching the other locomotive, the bird instantaneously turns around and heads back toward the first one. It continues flying in this manner. The question is, How far will the bird fly before being crushed between the two locomotives? If one concentrates on the distance the bird travels, the problem is difficult and requires one to add up the lengths of each lap flown. If, however, one looks at the time necessary for the trains to meet (2 hours, since they are traveling the 300 miles between them at a combined rate of 150 miles an hour), then it is easy to see that the bird travels $2 \times 200 = 400$ miles before being crushed. Actually, one locomotive derails at the last second and the bird is saved. This, of

3. Note the response of most people to these: "That's odd," or "You must be kidding."

course, has nothing to do with the "derailment" theory of humor.

An appropriate way to end a chapter entitled "Mathematics and Humor" is with some humor—more specifically jokes, since they are short and make sense without much context. The following jokes illustrate a few of the earlier-mentioned similarities between humor and mathematics.

First, a prototypical "dirty" joke: A fat, pompous man walks along, slips on a banana peel, and falls into a mud puddle.

Idiot and misunderstanding jokes usually are good illustrations of both superiority and incongruity theories of humor: Two idiots, one tall, skinny, and bald, the other short and fat, come out of a tavern. As they start toward home a bird flies over and defecates on the bald man's head. The short man says he's going back to the tavern for toilet paper, whereupon the tall one observes, "No, don't do that. The bird's probably a mile away by now."

A fat man (brother to the one in the previous joke) sits down to dinner with a whole meat loaf on his plate. His wife asks whether she should cut it into four or eight pieces. He replies, "Oh, four, I guess, I'm trying to lose weight."

A convict is playing cards with his guards. On discovering that he has been cheating, they kick him out of jail.

Finally, let us leave the jokes and go on to the mathematics with something of the mock earnestness expressed by Lewis Carroll in the following little poem:

> Yet what mean all such gaieties to me
> Whose life is full of indices and surds
> $X^2 + 7X + 53 = \frac{11}{3}$.

2

Axioms, Levels, and Iteration

As I mentioned in chapter 1, logic and deduction, besides playing an essential role in mathematics, are important to an understanding of humor. After all, one must have some grasp of logic even to recognize a nonsequitur. In addition to jokes that utilize the logical notions of reductio ad absurdum, presupposition, non sequitur, disguised equivalence, and so forth ("sillygisms" might be an appropriate term for such jokes), many jokes and riddles depend for their humor on an implicit understanding of the axiomatic method. We will see exactly how after I first develop a formal account of the axiomatic method and the distinction between object-level and metalevel statements. Later in the chapter I will also discuss the notion of iteration and its relevance to humor.

The axiomatic method goes back to ancient Greek geometry. Succinctly, it means selecting certain self-evident statements as axioms and deducing from them, by logic alone, other statements, which often are not so self-evident. This method is probably familiar from high-school geometry. What may not be so familiar is the idea that there may be different interpretations for a given set of axioms. This is possible since the axioms must contain undefined terms and since the deductions cannot depend on intuition about the subject matter of the axioms. (Regarding the first point, of course, other terms may be defined in terms of these undefined terms, but eventually one must accept certain terms as primitive.) The best way to get clear on these matters is to construct a simple example of an uninterpreted axiom system.

Thus we will first state the axioms very abstractly, then worry about what they might mean. The axioms for our simple axiom system follow; the letter *F* appearing in the axioms stands for an arbitrary relation between elements.

Axiom 1: If for any two things (elements) *a* and *b*, *b* stands in the relation *F* to *a*, then *a* does not stand in the relation *F* to *b*. (Abbreviated: if *b F a*, then not *a F b*.)

Axiom 2: For every element *a* there is an element *b* such that *b* stands in the relation *F* to *a*. (For all *a* there is a *b* such that *b F a*.)

Axiom 3: For every element *a* there is an element *b* such that *a* stands in the relation *F* to *b*. (For all *a* there is a *b* such that *a F b*.)

Axiom 4: For any three elements *a*, *b*, and *c*, if *b* stands in the relation *F* to *a*, and *c* stands in the relation *F* to *b*, then *c* stands in the relation *F* to *a*. (If *b F a* and *c F b*, then *c F a*.)

Axiom 5: For any two elements *a* and *b* such that *b* stands in the relation *F* to *a*, there is a third element *c* such that *c* stands in relation *F* to *a* and *b* stands in relation *F* to *c*. (If *b F a*, then there is a *c* such that *c F a* and *b F c*.)

What can we prove from these axioms? Since this is such a simple set of axioms, not much. Note, however, that we need not know what they are "about" in order to find their logical consequences.

Theorem 1: Given any element a, there exist infinitely many elements *b* such that *b* stands in relation *F* to *a*.

Given any element *a*, there exists at least one element *b* such that *b F a*. This follows from axiom 2. Now we can apply the axiom again, this time to element *b*. Thus there is some element, call it *c*, such that *c F b*, again by axiom 2. But if *c F b* and *b F a*, then by axiom 4, *c F a*. It is clear that we can repeat this process of applying axioms 2 and 4 indefinitely. Thus we conclude that there are infinitely many elements standing in relation F to a.

We can also prove the following similarly unremarkable result.

Theorem 2: Given any two elements a and b such that $a\,F\,b$, there exist infinitely many elements c having the property that $a\,F\,c$ and $c\,F\,b$.

If $a\,F\,b$, we know from axiom 5 that there is at least one element c such that $a\,F\,c$ and $c\,F\,b$. Since $a\,F\,c$, there is, by axiom 5 again, an element d such that $a\,F\,d$ and $d\,F\,c$. Now, by axiom 4, since $d\,F\,c$ and $c\,F\,b$, we know that $d\,F\,b$. Thus $a\,F\,d$ and $d\,F\,b$, and we have a second element with the required property. We can continue to use axioms 5 and 4 in succession indefinitely to find elements p such that $a\,F\,p$ and $p\,F\,b$.

Now that we have stated the axioms and proved some simple theorems, we should ask again what these axioms are about. One interpretation that probably occurred to you is to understand the elements to be points on a line and F to be the relation "to the right of." On this interpretation the axioms are self-evidently true (see fig. 5).

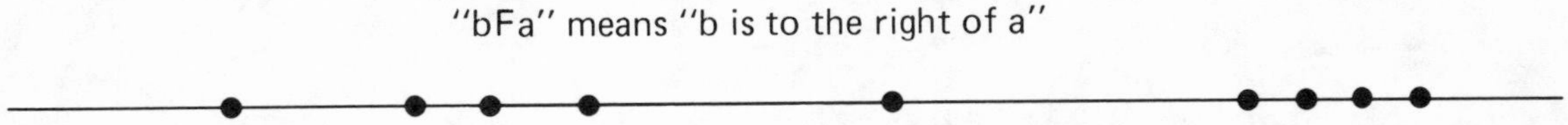

"bFa" means "b is to the right of a"

Fig. 5

Axiom 1 thus says that if b is to the right of a, then a is not to the right of b. Axioms 2 and 3 say that there are no end points on either side of the line. Axiom 4 states that if b is to the right of a, and c is to the right of b, then c is to the right of a. Relations like "to the right of," which have this last property, are called transitive. Finally, axiom 5 says that the relation is "dense"; that is, between any two points there is a third point.

An interpretation that gives a meaning to the abstract elements and to the undefined relation symbol F and in which the axioms come out true is called a *model* of the axioms. Thus the points on a line with "to the right of" interpreting F are a model for axioms 1 to 5. Are there any essentially different models of these axioms? Since I introduced this example to illustrate that a set of axioms can have more than one interpretation, it is not hard to guess that there are. (One may think of the axioms as being clues to a mystery and the different possible scenarios for the crime as being models of these axioms.)

For example, consider the elements to be all the circles on a plane surface and F to be the relation "is contained within." Then axiom 1 says that if circle b is contained in circle a, then a is not contained in b (see fig. 6). Axioms 2 and 3 state that given any circle there is another one containing it as well as one contained in it. Axiom 4 says that if b is contained within a and c within b, then c is within a; axiom 5 says that if one circle contains another, there is a third circle within the first that contains the second. Thus the collection of all circles in the plane with F interpreted as "is contained within" is also a model for axioms 1 to 5.

The theorems proved from the axioms by logic above hold true for *all* models, hence for this one in particular. This is

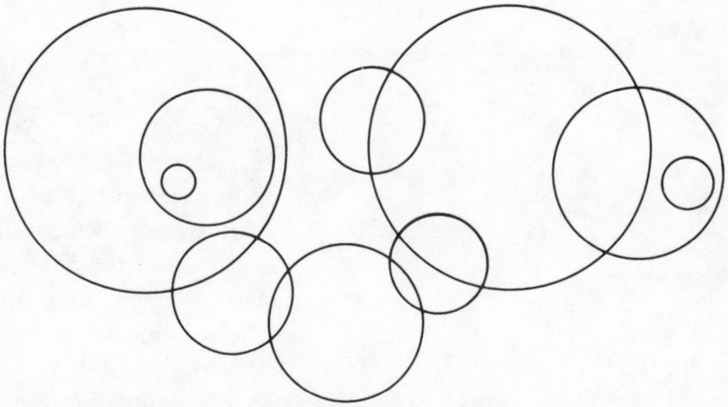

Fig. 6 "bFa" means "b is contained within a"

worth repeating. If a theorem follows from a set of axioms by logical deduction, then that theorem must hold for all models of the axioms. This fact gives us a method for determining when a statement is *not* provable from the axioms. If a statement can be seen to be true in some models of the axioms and false in others, then that statement cannot be proved from the axioms. There is nothing deep here—just a slightly more precise account of the commonsense idea that a statement cannot be proved if it has a counterexample.

Consider now statement *S*: Given any element *a*, there is a *b* (different from *a*) such that neither *b F a* nor *a F b*. *S* is false in our first model of axioms 1 to 5, since there it says that there is a point *b* on the line different from *a* that is neither to the right of *a* nor to the left of it. *S* is, however, true in the second model of the axioms, since there it says that given any circle *a* there is another circle *b* that neither contains *a* nor is contained in *a*. Figure 6 illustrates the truth of *S* in the second model. Thus *S*, which is false in one model of the axioms and true in another, can be neither proved nor disproved from the axioms. Such a statement is said to be *independent* of the axioms. I will refer again to independent statements later in this chapter and at the end of chapter 3.

Before returning to humor, let me enunciate an extremely important distinction both in mathematical logic and in discussions of humor—the distinction between the object level and the metalevel. Object-level statements are statements *within* the axiom system being studied. Examples are:

i) If *a F b* and *b F c*, then *a F c*.
ii) *a F b* or *b F a* or $a = b$.
iii) For all *b*, *b F a*.
iv) There is a *b* such that *b F a*.

Metalevel statements are statements *about* the axiom system or about the object-level statements within it. Examples are:

i) *S* is independent of axioms 1–5.
ii) Axioms 1–5 have two different interpretations.

iii) *S* is true in an interpretation.

iv) Axiom 5 is more interesting than axiom 1.

What does this all have to do with humor? In the first chapter I stated that a necessary ingredient for humor is that two incongruous ways of viewing something (a person, a statement, a situation) be juxtaposed; that is, for something to be funny, some unusual, odd, or inappropriate aspects of it must be seen or imagined together and compared. Axiom systems and their interpretations or models provide a formal analogue for a certain sort of incongruity, namely that resulting from a statement or story having two different and incongruous interpretations. Moreover, since the two incongruous interpretations both satisfy the same statement or story, there is some point to the incongruity as well.

The formal structure of such stories or jokes is as follows. Joke-teller: "In what model are axioms 1, 2, and 3 true?" Listener: "In model M." Joke-teller: "No, in model N." The following classic burlesque joke is an example (fig. 7). The dirty old man leers at the innocent young virgin and says, "What goes in hard and dry and comes out soft and wet?" The girl blushes and stammers, "Well, let's see, uh . . . ," to which the dirty old man replies wickedly, "chewing gum." In other words, "model N" in our formal example and "chewing gum" (more accurately the whole scenario suggested by chewing gum) in our burlesque joke play the role of an unexpected and incongruous model of the given "axioms." Note the similarity to the Freudian joke of chapter 1.

Needless to say, in most jokes of this type the "axioms" are implicit and are expressed only in abbreviated, elliptical, and colloquial terms. The "natural" interpretation for the axioms is a familiar one. The punch line of the story provides some other unexpected and incongruous interpretation that, if the mood is right, results in humor. Consider as a second example the story of the young man who registered his requirements at a computer dating service. He wanted someone

Fig. 7

who enjoyed water sports, liked company, was comfortable in formal attire, and was very short. The computer sent him a penguin. It is clear that the young man's requirements play the role of axioms and that the natural interpretation of these axioms is a young woman with a life-style meeting these requirements. The penguin and its life-style provide the axioms with an unexpected model.

Riddles also have the same formal structure as the type of joke just discussed. "What has properties A1, A2, and A3?" "M" (or sometimes "I don't know"). "No, N." Homonyms often play a role in riddles as well. Consider the very common riddle: "What's black and white and red all over?" "A newspaper." Frequently, of course, there is more than one incongruous interpretation for a riddle. M. E. Barrick (1974) has compiled a monstrously long list of answers to the above riddle that includes: a wounded nun; an embarrassed zebra; Santa Claus coming down a dirty chimney; a right-winger's view of an integration march; and a skunk with diaper rash.

My favorite joke-riddle of this sort (it is actually more a parody) appears in Leo Rosten's *The Joys of Yiddish* (1968). A father asks his son, "What is it that hangs on the wall, is green, wet, and whistles?" The boy thinks for a while and, perplexed, finally gives up. "A herring," the father says. "A herring? A herring doesn't hang on the wall," the son points out. "So hang it there," the father reasons. "But a herring isn't green." "So paint it." "But a herring isn't wet." "If it's just painted, it's still wet." "But," the exasperated son sputters, "a herring doesn't whistle." "Right," smiles the father. "I just put that in to make it hard."

I should emphasize here that to get (i.e., understand) a joke, either situational or canned, one must ascend, so to speak, to the metalevel at which both interpretations, the familiar and the incongruous, can be imagined and compared (or, if there is only one interpretation, at which its oddness can be appreciated). This seems clear for the type of joke just discussed and will be seen to be true for other types as well. The various interpretations and their incongruity of course depend critically on the context, the prior experience of the person(s) involved, their values, beliefs, and so on.

The necessity of this psychic stepping back (or up) to the metalevel is probably what is meant when people say that a sense of perspective is needed for an appreciation of humor. It also explains why dogmatists, idealogues, and others with one-track minds are often notoriously humorless. People whose lives are dominated by one system or one set of rules are stuck, in a manner of speaking, in the object level of their system. Whether they are political radicals mouthing some party line or bureaucrats blindly enforcing some set of petty regulations, they lack the ability to step outside themselves and their systems. Understanding a joke is a distinctly human activity and requires one to evaluate almost instantly the relative importance of its different parts, to compare meanings and shades of meaning, to perceive unstated relations and implicit ideas, and to put this all into an appropriate context in order to grasp the situation as a whole. These complex operations are all metalevel (or meta-metalevel) activities and are beyond the capabilities of computers and people who want to be computers.

Nevertheless, the rigidity of such people is sometimes itself unintentionally funny (if they do not have power over you). The incongruity of a human being behaving as an automaton is probably the reason. Bergson would most likely agree.

At the other extreme from these would-be automatons we find people whose minds are mush (in the sense of being extremely loose and unstructured). Such people are not likely to have much sense of humor either. This is so because a modicum of mental orderliness, the awareness of various complexes of ideas and their links to one another, and the (at least partial) acceptance of certain values is necessary to an appreciation of humor. With no feeling for what is correct, congruous, or natural, there can be no perception of what is incorrect, incongruous, or unnatural.

A somewhat similar idea was expressed by the anthropologist Ralph Piddington (1933) when he stressed that laughter presupposes a system of values and beliefs. The French sociologist Dupréel (1928) was one of the first to comment that shared laughter often reinforces such values and beliefs (axioms, in our formal terms) and seems to demarcate social groups. More recently Lawrence La Fave, a psychologist (1978), has found empirical support for the statement that a joke is humorous to the extent that it enhances a "positive reference group" or disparages a "negative reference group." Thus a joke (more accurately an "insult" joke) is funny when, as La Fave writes, the good guys (positive reference group) win and the bad guys (negative reference group) lose and is not funny when the bad guys win and the good guys lose. This relativity of humor holds in general and depends on the simple facts that the notions of incongruity and inappropriateness derive from the prior notions of congruity and appropriateness and that different people (different reference groups) have different standards (axioms) for what is congruous and appropriate. I will return to this at the conclusion of the book.

I do not want to leave the topic of alternate models for axiom systems without saying something about non-Euclidean geometry. Euclid's axiomatic development of geometry is no doubt familiar. Among these well-known axioms is the famous parallel postulate, which says that, through a point not on a given straight line, we can draw exactly one straight line parallel to the given line. Here *point* and *straight line* are undefined terms, and two straight lines are defined as parallel if they have no point in common. Interpreting point and straight line in the usual way gives us the diagram in figure 8.

Many mathematicians through the centuries tried to prove the parallel postulate (axiom) from the other axioms of geometry. They used every method imaginable to them including reductio ad absurdum but could never come up with a proof. This failure seemed to give Euclidean geometry a certain absoluteness. Immanuel Kant even claimed that people could think about space only in Euclidean terms. Finally, in the nineteenth century the mathematicians Gauss, Bolyai, and Lobachevski realized that Euclid's parallel postulate

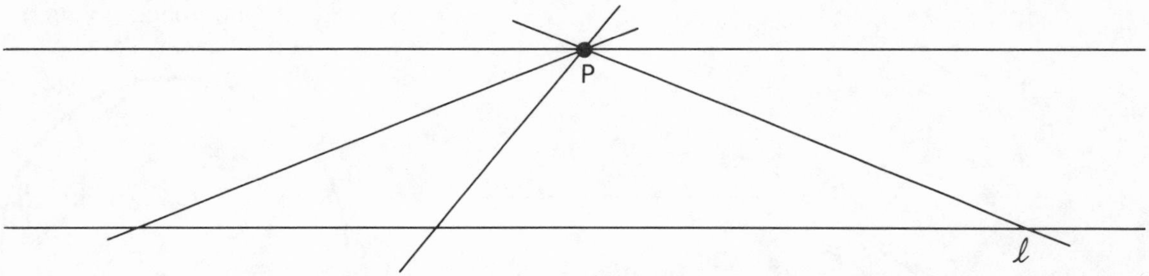

Fig. 8

bore the same relation to the other axioms of Euclidean geometry as statement *S* in our little formal system bore to axioms 1–5; in other words, it was independent of them. Later a model of the other axioms of Euclidean geometry in which the parallel postulate was false was constructed. The notion of different interpretations for axiom systems was not known before this time.

Though admittedly straining the meaning of the word *joke*, the discovery of another interpretation for Euclid's axioms (without the parallel postulate included) is a sort of mathematical joke.[1] (It is a joke that Immanuel Kant did not get.) The emotional climate mentioned in chapter 1 is not quite right here, but there is a sort of intellectual smile, even if not howls of laughter, associated with recognizing the structure we are about to develop as also being a model for Euclid's axioms (without the parallel postulate included).

1. Similar statements can be made about Abraham Robinson's nonstandard models for analysis.

In our new model (due to Poincaré) the Euclidean plane is replaced by a fixed circle, points (an undefined term) in the plane are interpreted by points within this circle, and straight lines (another undefined term) in the plane are interpreted by circular arcs that cut the circle perpendicularly or by diameters of a circle (see fig. 9).

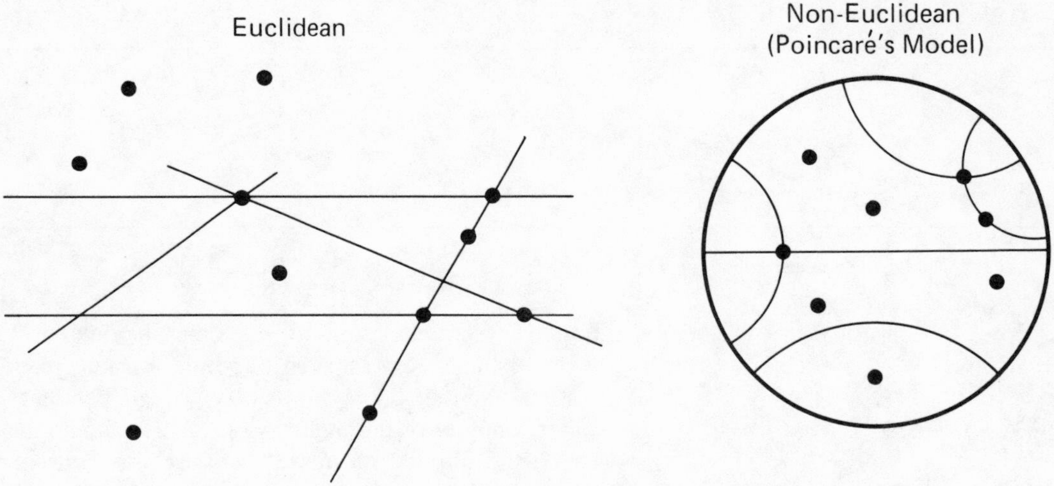

Fig. 9

Distance is defined in such a way that intervals near the circumference of the fixed circle are longer than those near the center. In fact the length of a "straight line" (i.e., a circular arc cutting the fixed circle perpendicularly) is infinite, since an interval on such an arc can be made arbitrarily long by moving it toward the circumference of the fixed circle (fig. 10).

Given this understanding of the basic undefined terms—points, straight lines, and distance—we can check that all the axioms of Euclidean geometry with the exception of the parallel postulate are true in this interpretation.

For example, it is not hard to see that through any two points there is one "straight line" (fig. 11*a*). Owing to the

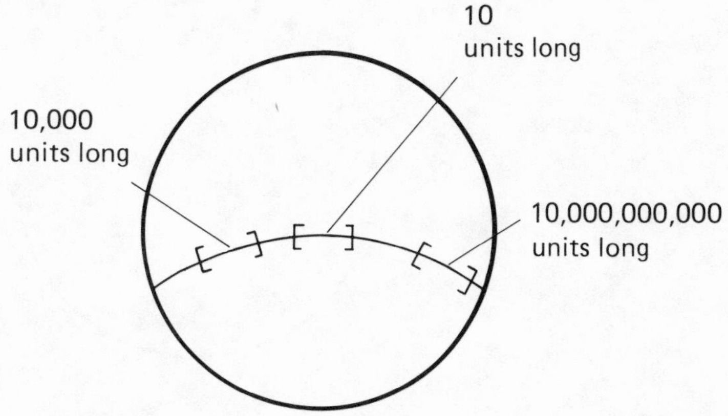

Fig. 10

way distance is defined, any line segment can be indefinitely extended. Furthermore, circles can be drawn about any point as center yet look somewhat elliptical because of the way distance is defined (fig. 11*b*).

Finally, let us check that the parallel postulate is false in this interpretation. It is easy to see (fig. 12) that through a point *p*, more than one line (actually, infinitely many lines) can be drawn parallel to a given line *l*. The parallel postulate is then seen to be independent of the other axioms of geometry. It is true in some models of these axioms and false in others (namely, this one); hence it cannot be proved from these axioms.

There are other models in which the parallel postulate is false. (In terms of our humor analogy, Euclidean geometry without the parallel postulate is thus a very good joke.) Axiom systems that incorporate the denial of the parallel postulate as an axiom are called non-Euclidean geometries. Which geometry is true of the real world seems to be partly a matter of convention and partly an empirical question. Einstein found it convenient to assume that space is non-Euclidean (but unlike the non-Euclidean model considered above).

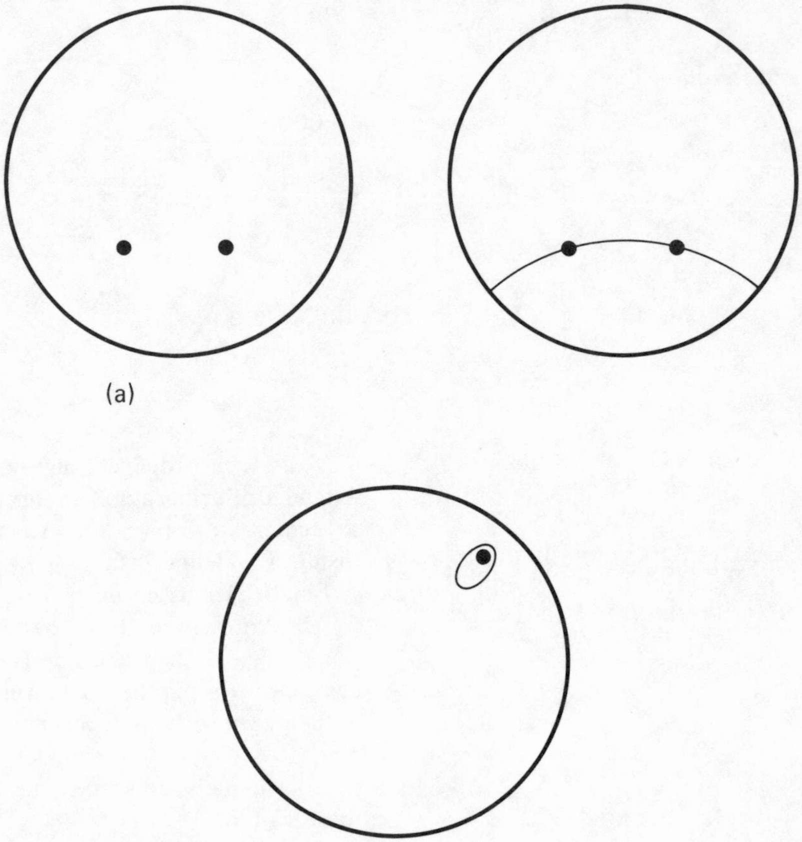

(a)

(b)

Fig. 11

I will end this chapter with a brief discussion of an operation important in humor, mathematics, and computer science—the operation of iteration. Counting—adding 1 to previous integers—is probably the simplest and most important example: 1, 2 = 1 + 1, 3 = 2 + 1, 4 = 3 + 1, . . . The great French mathematician Poincaré considered the

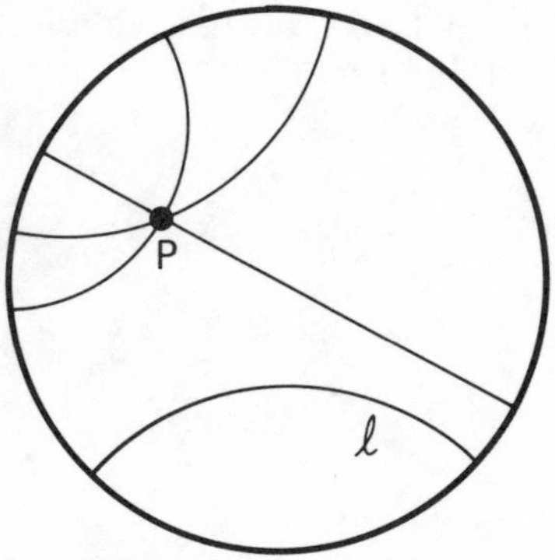

Fig. 12

natural numbers (1, 2, 3, . . .) and iteration to be at the base of all mathematics. Before discussing its relevance to humor, let us consider a few examples of the use of iteration in mathematics (and computer science). The details of these examples are not essential to the sequel.

Addition and multiplication can be defined in terms of counting (adding 1) by means of certain iteration procedures. Thus, to add x to y, simply add 1 to the sum of x and the predecessor of y. But, to find the sum of x and the predecessor of y, it is necessary to add 1 to the sum of x and the predecessor of the predecessor of y. This process continues until 0 is reached. The sum of x and 0 is defined to be x. How to add 4 to 5 is summarized by the following equations:

$$5 + 4 = (5 + 3) + 1;$$
$$5 + 3 = (5 + 2) + 1;$$
$$5 + 2 = (5 + 1) + 1;$$
$$5 + 1 = (5 + 0) + 1.$$

Hence, $5 + 4 = ((((5 + 0) + 1) + 1) + 1) + 1)$.

Similarly, to multiply x by y, multiply x by the predecessor of y and add x. But, to multiply x by the predecessor of y, it is necessary to multiply x by the predecessor of the predecessor of y and add x. This process is continued until 0 is reached; x multiplied by 0 is defined to be 0. How to multiply 5 by 4 is summarized by the following equations:

$$5 \times 4 = (5 \times 3) + 5;$$
$$5 \times 3 = (5 \times 2) + 5;$$
$$5 \times 2 = (5 \times 1) + 5;$$
$$5 \times 1 = (5 \times 0) + 5.$$

Thus $5 \times 4 = ((((5 \times 0) + 5) + 5) + 5) + 5)$. Since addition is defined in terms of counting by iteration, and since multiplication is defined in terms of adding by iteration, multiplication can also be defined in terms of counting by iteration. These sorts of considerations and others like them give some plausibility to Poincaré's statement above.

(There is a version of Poincaré's statement, "Church's thesis," that is considerably more plausible. See Rogers [1967] for details.)

A somewhat different geometric example shows something of the power of the method. The problem is to find where a curve crosses a given line, and the solution by successive approximating iterations is due to Isaac Newton. Consider the curve below crossing line *l* at point *P* (fig. 13).

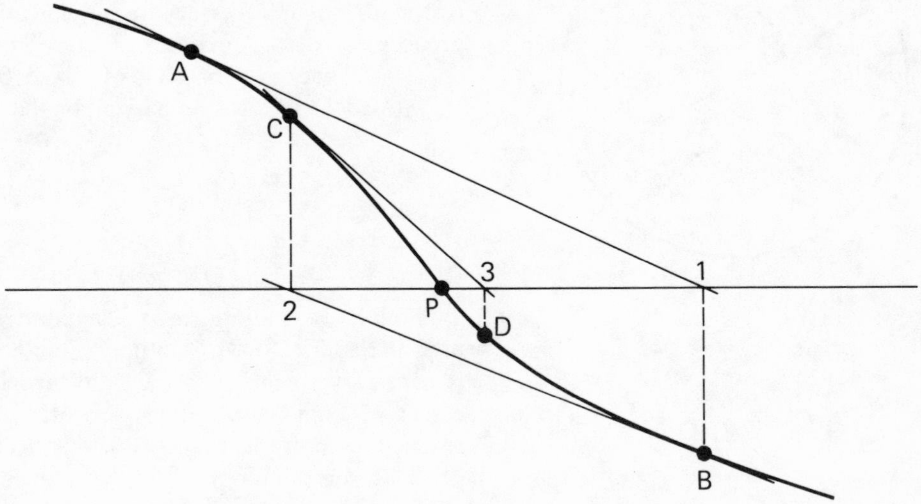

Fig. 13

An arbitrary point *A* on the curve is chosen, and the line tangent to the curve at that point is drawn and intersects line *l* at point 1. Then point *B* on the curve is located by drawing a perpendicular line from 1 to the curve. Now the process is iterated and a line tangent to the curve at *B* is drawn and intersects line *l* at point 2. Point *C* on the curve is located by drawing a perpendicular line from 2 to the curve. Iterate again and point 3, the third approximation to *P*, is reached. By continuing this process we can get as close to point *P* as we want.

Our last example concerns the notion of a mathematical function, a notion that will be needed in chapter 5. A mathematical function is a rule that establishes a correspondence between pairs of elements (numbers, in our case). Thus "$f(x) = 2x$" is a rule that associates with any number x another number twice as large, namely $2x$. Hence the function f associates with 3 the number 6, with 4 1/4 the number 8 1/2. This is indicated by "$f(3) = 6$" and "$f(4\ 1/4) = 8\ 1/2$." Similarly, "$g(x) = 3x^2 - 1$" associates with any number x a number three times its square minus 1. Thus $g(2) = 11$ and $g(3) = 26$. Likewise, if $h(x) = x^2 + x = 6$, then $h(1) = 8$ and $h(3) = 18$.

Functions (rules) can be iterated, and this operation, called composition of functions, is very common in mathematics. Take an arbitrary function, say $g(x) = 3x^2 - 1$, and an arbitrary number, say 1, and find $g(1) = 2$. Then find $g(2) = 11$. Iterating, we find $g(11) = 362$, and so on. Or consider $p(x) = x^2$ and choose the initial number to be 1/2. Then $p(1/2) = 1/4$, $p(1/4) = 1/16$, $p(1/16) = 1/256$. Often iteration of functions has interesting geometric or physical interpretations.

The functions we will need in chapter 5 are those that establish a correspondence among triples of numbers; more accurately, they are rules that associate with any pair of numbers a third number. For example, "$z = f(x,y) = x + y$" is a rule that associates with any pair of numbers another number, nearly its sum. Thus, $f(6,5) = 11$, and $f(2,-5\ 1/2) = -3\ 1/2$. And "$z = g(x,y) = x^2y - yx^3$" associates with any pair of numbers the square of the first times the second minus the second times the cube of the second. Hence $g(2,3) = 2^2 \times 3 - 3 \times 2^3 = -12$ and $g(1,5) = 1^2 \times 5 - 5 \times 1^3 = 0$.

What, finally, does iteration have to do with humor? This operation is an important factor in many jokes and humorous situations. It is the mechanical and repetitive carrying out of some formula or algorithm—and, as Bergson wrote, repetitive or mechanical actions are the essence of humor, since they violate the characteristic flexibility and spirituality of human beings. In fact, puppets and jack-in-the-boxes were cited by Bergson as primary examples of humorous rigidity (rule-determined behavior by humans or humanlike things). Idiot or ethnic jokes wherein the idiot or the ethnic person repeatedly and blindly follows some rule or inappropriate convention are another example.

More generally, it is well known that repeated display of character traits or mannerisms often is the key to a comedic personality. Certain stock types such as the pompous braggart go back to Aristophanes. Jack Benny's cheapness, W. C. Fields's misanthropy, and Charlie Chaplin's walk are more contemporary examples of the same phenomenon. Most well-known comedians, in fact, develop a persona that is to a certain extent stylized, repetitive, and predictable.

Comedies as well as comedians depend for part of their humor on mere repetition (iteration, if you will). The critic Northrop Frye has commented that even tragic events repeatedly enacted begin to become funny. Parents losing their child to measles, say, and their consequent grief and suffering are very sad. But if a play were to depict the death from measles each year for seven years of one of their seven children and the parents' consequent grief, the tragedy would soon turn to comedy (of a measly sort). A related phenomenon holds in comic strips where the characters begin to be really funny only after their identities as gluttons, morons, shrews, or such have been established through repetition day after day. Television situation comedies also usually are funnier to regular viewers who are aware through repetition of the main characters' quirks.

There are various devices that make such repetition possible, and even a casual acquaintance with comedies from Shakespeare to Neil Simon will let us recognize them. Mistaken identity and role reversal are very common examples enabling the playwright (or the author in general) to capi-

STAN'S TOWING
STAN'S TOWING
JIM'S TOWING
STAN'S TOWING
JIM'S TOWING
JOE'S TOWING

Fig. 14

talize repeatedly on the resulting incongruity. Placing a character in a strange land or culture likewise permits the writer to exploit the situation repeatedly. Other factors are of course involved, but iteration, simple though it is, is an important component of comedy.

Many jokes use repetition for emphasis or to establish the correct rhythm. In certain kinds of jokes, though, it plays a more important role. "Shaggy-dog" stories are narratives that are indefinitely prolonged. Innumerable episodes, all of the same general kind, are included until finally a punchless non sequitur of a punch line is reached. They are popular among children, who often pounce on the teller when he finishes.

The element of iteration is apparent in children's play as well. Play, though not usually called humor (it lacks a punch line, for instance), is certainly closely related to humor. Games like pat-a-cake, peek-a-boo, tag, jumping rope, Simon says, and hide-and-seek all involve a simple rule that is iterated repeatedly.

Iteration is often combined with some form of self-reference, as in figure 14. Self-reference (and paradoxes resulting from it) will be the main topic of the next chapter.

3

Self-Reference and Paradox

The notion of self-reference is at the root of a wide class of jokes and some famous paradoxes and theorems in mathematical logic, and it is crucial to an understanding of humor in general. I will begin by considering a classic paradox. It concerns Epimenides the Cretan, who stated that all Cretans are liars (hence the term "self-reference" in the title). The crux of the paradox is clearer if we simplify his statement to "I am lying" or, better yet, "This sentence is false."

Let us give the label Q to "This sentence is false." Now we notice that, if Q is true, then by what it says it must be false. On the other hand, if Q is false, then what it says is true, and Q must then be true. Hence, Q is true if and only if it is false.

A different but related paradox concerns the barber of Seville. He was the only barber in Seville, and he was reportedly ordered by law to shave all those men and only those men who did not shave themselves. The paradoxical nature of the order is apparent when we ask who shaves the barber. If he shaves himself, by law he should not. On the other hand, if he does not shave himself, by law he should. Requiring that the barber be nine years old is cheating.

Another version of this paradox (of which there are many) concerns the mayors of cities in a certain country. Some of the mayors live in the cities they govern, others are nonresident mayors. A law is passed requiring all and only the nonresident mayors to live in one place—call it city C. City C requires a mayor. Where shall the mayor of city C reside?

There is a close connection between these paradoxical laws and "double bind" situations. The simplest such situation is generated by the command "Be spontaneous." Most situations that require contradictory behaviors are somewhat disguised, however, and are therefore more insidious. In fact, the philosopher Saul Kripke (1975) has observed that two or more nonparadoxical sentences may, taken together, yield a liar paradox or a double bind. The psychiatrist R. D. Laing (1970), among others, has done some interesting work on the behavioral consequences of this.

Before returning to a further discussion of paradoxes, however, let us examine some humorous examples of self-contradictory self-reference. Modal jokes result when the content of a statement is incongruous with its form or mode of expression. That is, the statement's mode of expression belies its content, and the resulting incongruity is often humorous. The billboard advertisement in figure 15 is an example.

Other examples are desk plates with "Plan Ahead" squeezed onto them, lapel buttons with the message "Support Mental Health or I'll Kill You," or a hysteric screaming "Relax!"

This type of humor is actually very pervasive. Almost any kind of presentation can be made humorous by making it incongruous with its content. A symphonic treatment (by Mahler, for example) of "For He's a Jolly Good Fellow" or a rock version of *Madama Butterfly* are thus in a sense modal jokes. The same is true of an epic portrayal of the local weather report or a cartoon version of *Gone with the Wind*.

Fig. 15

Modal jokes of course have much in common with simple irony, one of my favorite examples of which is the following true story.

A well-known, but here anonymous, philosopher was delivering a talk on linguistics and had just stated that the double negative construction in some languages has a positive meaning and in some a (very) negative meaning. He went on to observe, however, that in no language was it the case that a double positive construction has a negative meaning. To this another well-known philosopher in the rear of the lecture room responded with a jeering "yeah, yeah."

Related to modal jokes are Russell jokes, jokes whose logical underpinning is some version of Russell's paradox or its resolution (a topic I will get to presently). These jokes involve iteration and self-reference. A neurotic's worrying about not having any worries is an example, as is the man who, consciously moderate in all facets of his life, suddenly realizes he has been immoderately moderate. Similarly, phrases like "bored with boredom," "tired of being tired," "anxious about my anxieties," manifest the same phenomenon, as do the following interchange and figure 16. Young man: "Why do philosophers ask so many questions?'" Old philosopher: "Why shouldn't philosophers ask so many questions?"

Russell's paradox is stated in terms of set theory; it is an abstract version of the barber and mayor paradoxes. Informally, a set is a collection of objects of any sort whatever. Examples of sets thus are (1) the collection of faculty members of Temple University during fall semester 1977; (2) the collection of prime numbers; (3) the collection of cantaloupes in Nairobi on 8 July 1961; and (4) the collection of functions from the whole numbers to the whole numbers. Professor Alu Srinivasan is a member of the first set; 6 is not a member of the second set; a watermelon is not a member of the third set (neither is Alu Srinivasan); and the function f defined on whole numbers such that $f(x) = 2x$ is a member of the fourth set.

Set theory is a beautiful subject full of ingenious arguments and surprising counterexamples and should be mastered by anyone interested in mathematics and its foundations. All we need to derive Russell's paradox, however, are a few of the following elementary definitions and notations.

"$x \in y$" means that x is a member of the set y. If y is the set of countries in the United Nations and x is Brazil, then $x \in y$.

WHAT DO YOU THINK IS THE CAUSE OF THE IGNORANCE AND APATHY OF THE VOTING PUBLIC?
I DON'T KNOW AND I DON'T CARE.

Fig. 16

"$x \notin y$" means that x is not a member of the set y.

"$x \subset y$" means that x is a subset of y; that is, every member of x is also a member of y. If y is as above and x is the set of North American countries, then $x \subset y$. If x is not a subset of y, we write $x \not\subset y$.

There are three basic operations defined on sets. "$x \cap y$" refers to that set whose members belong to *both x and y*. It is read "x intersection y."

"$x \cup y$" refers to that set whose members belong to x *or* to y *or* to both. It is read "x union y."

"$\bar{x}$" refers to that set whose members do *not* belong to x. It is read "x complement."

Usually $\bar{x}$ refers to that set whose members do not belong to x but do belong to some other relevant set. A set is often indicated by listing its elements within brackets. $x = \{$Sheila, Leah, Daniel$\}$ is thus a set having three members—Sheila, Leah, and Daniel.

As an illustration, let $x = \{2, 4, 5, 7, 8\}$ and $y = \{1, 2, 4, 7, 9\}$. Then $x \cap y = \{2, 4, 7\}$ and $x \cup y = \{1, 2, 4, 5, 7, 8, 9\}$. $\bar{x} =$ the set whose elements do not belong to x. Thus, $3 \in \bar{x}$, $41{,}283 \in \bar{x}$, and Mark Twain $\in \bar{x}$. In this context it is more natural to take $\bar{x}$ to be the set whose elements do not belong to x but do belong to $z = \{1, 2, 3, 4, 5, 6, 7, 8, 9, 10\}$. In this case $\bar{x} = \{1, 3, 6, 9, 10\}$. Finally, if we define q as $q = \{2, 4, 9\}$, then $q \subset y$.

Returning to the derivation of the paradox, we note that some sets contain themselves as members (symbolically, $x \in x$). The set of all things mentioned on this page is mentioned on this page and thus contains itself. Likewise the set of all those sets with more than seven members itself contains more than seven members and thus is a member of itself. Most naturally occurring sets do not contain themselves as members (symbolically, $x \notin x$). The sets of hairs on my head on 6 May 1977 is not itself a hair and thus is not a member of itself. Similarly, the set of odd numbers is not itself an odd number and thus does not contain itself as a member.

Dividing the set of all sets into two nonoverlapping sets, let us denote by M the set of all those sets that do contain themselves as members and by N the set of all those sets that do not contain themselves as members. In other terms, for *any* set x, if $x \in M$, then $x \in x$; conversely, if $x \in x$, then $x \in M$. On the other hand, for *any* set x, if $x \in N$, then $x \notin x$; conversely, if $x \notin x$, then $x \in N$. Now we may ask whether N is a member of itself or not. (Compare this question with "Who shaves the barber?" and "Where does the mayor of city C live?") If $N \in N$, then by definition $N \notin N$. But if $N \notin N$, then by definition $N \in N$. Thus N is a member of itself if and only if it is not a member itself. This contradiction constitutes Russell's paradox.

A resolution of this paradox is to restrict the notion of a set to a well-defined collection of already existing sets. An axiomatic set theory was developed that formalizes the accepted principles of set theory and excludes (it is hoped) "bad" sets like M and N. Bertrand Russell in his famous theory of types (1910) classified sets according to their type or level. On the lowest level, type 1, are individual objects. On the next level, type 2, are sets of type 1 objects. On the next level, type 3, are sets of sets of type 1 or of type 2, and so on. The elements of type n sets are sets of type $(n - 1)$ or lower. In this way Russell's paradox is avoided, since a set can be a member only of a set of a higher type and not of itself. A set's being a member of itself ($x \in x$) is thus ruled out, as are sets like M defined in terms of this notion.

I should mention that Russell and Whitehead constructed the theory of types not only to prevent the paradox (and others like it) but, more importantly, to provide an axiomatic foundation for the whole of mathematics. They succeeded in reducing all of mathematics to logic as embodied in the theory of types (logic together with the above hierarchical notion of set).

Note that in Russell's resolution of the paradox we are once again led to the notion of levels. In chapter 2 I discussed the distinction between object-level statements within the formal system and metalevel statements about the formal system. In Russell and Whitehead's theory of types we have different levels within the formal system itself and a metalevel in which we talk about all the object levels (types).

Applying a type solution to the Cretan paradox requires that "All Cretans are liars" be assigned a higher type number than other statements made by Cretans. We must make a distinction between first-level statements (usually called first-order statements), which do not refer to other statements at all; second-order statements, which refer to first-order statements; third-order statements, which refer to second-order statements; and so on. Thus, if Epimenides the Cretan states that all statements made by him are false, he is to be understood as making a second-order statement that does not apply to itself but applies only to first-order statements. Or he may assert that all his second-order statements are false. This assertion would then be a third-order statement and thus again would not apply to itself. In this way the self-reference of the Cretan paradox is prevented. More generally, the whole concept of truth is given a level structure: truth_1 for first-order statements, truth_2 for second-order statements, and so forth. This notion of truth has been extensively developed by the logician Alfred Tarski (1936).

As we have seen in the case of Russell jokes, this level structure of statements is often used humorously. Note also the practice common among comedians of making a comment (metastatement) on jokes that fail, thereby sometimes salvaging a metajoke. More generally, the ability to make self-deprecating remarks requires that one be able to view (a part of) oneself from a more neutral (meta-) vantage point.

It is increasingly common in modern literature and movies for there to be an alternation between the object level and the metalevel. The movies of Mel Brooks and Woody Allen, for example, contain many instances of the characters' stepping out of the story, commenting on it, interacting on the metalevel (even on the meta-metalevel and beyond), then reentering the story. The more frequent use of the practice recently is probably due to an increased self-consciousness and a keener taste for abstractness and paradox. Nevertheless, it is a very old idea. The chorus in classical Greek theater (along with its various descendants and offshoots through the Middle Ages, Shakespeare, etc.) was a kind of institutionalized commentator (metalevel) that also played an essential part on the object level of the play.

A complex interplay between levels has a role in many jokes and humorous situations. The following old joke is an example where this factor is simple and isolated and therefore clearer.

A joker says to an acquaintance, "Did you hear the one about the bride with gas? As the bride walks up the aisle, she suddenly leans over to her father and whispers, 'Daddy, the gas is terrible. I can't stand it anymore. What am I going to do?' Her father says, 'Wait till we get near the roses.' " Then the joker stops, bends closer, and says anxiously, "Did you hear it? Did you hear it?" The acquaintance, thinking he is being asked if he has heard the joke before, answers no. At this the joker says, "Neither did I. I was in the back of the church."

There are a couple of funny aspects of this joke, but the one that is important here is that the acquaintance misinterprets "Did you hear it?" to be a metalevel question about the joke and not an object-level question that is part of the joke.

Until now the word *level* has been used only in the expressions *object level* and *metalevel.* There are, of course, other senses of level, important to the meaning of a statement, that have nothing to do with this distinction. Thus we sometimes speak of the emotional level of a statement (or question, interjection, etc.). Or in poetry one sometimes speaks of the content's being reinforced on another dimension or level, say by sound or rhythm. Another common use occurs when we say that a story has many levels. Usually we mean not only meta- and object levels but more coordinate sorts of levels—a simple story, an allegory, an interesting adventure, a reply to something or someone, and so on. There are also other informal uses of level to mean the degree or extent of some attribute, the type of language appropriate for a given situation, and so on.

These different notions of level can combine, intertwine, or clash to give texture and depth to a (humorous) story. They cannot be simply ordered as to importance. By this I mean one cannot say that one sense of level is always more important than another, nor can one say, given a particular sense of level—say, emotional level—that one emotional state is always "higher" than another one. The ordering of these levels is very complex and partial,[1] as is the ordering within them.

Humor, let me reiterate, though it may use formal devices, depends ultimately on one's sensitivity to the interplay among the various "levels" of meaning. It is a very complex skill, this ability to distinguish levels of meaning, perceive their relationship, evaluate their relative importance given the context, then almost simultaneously form a global impression. Appreciating humor—even recognizing it—requires human skills of the highest order (level?); no computer comes close to having them.

A. M. Turing, the first major theoretician of computer science, once declared (1950) that the question whether

1. The mathematical structure known as a "partial ordering" captures the notion of incompatibility to which I am referring. Intuitively, a partial order is any set with an ordering in which, of any two elements, one is not always greater than the other. Two elements may simply be incomparable with respect to the given ordering. Most humanly interesting properties—beauty, intelligence, or wealth, for example—are less simplistically discussed in terms of partial rather than total orderings.

computers could ever be considered conscious was too vague to answer. He proposed replacing it with the more precise question whether a computer could be programmed to "fool" a person into believing he was dealing with another human instead of a computer. The person could pose yes-or-no or multiple-choice questions to the computer and to another person, both hidden behind a screen, and would then have to decide which set of answers came from the computer and which from the human. (This question can be refined in many other ways that need not concern us here.) This second question seems clearly answerable from our discussion of levels, context, and so forth: no computer (certainly no present-day computer) could disguise its inhumanity. All that would be necessary would be to ask it to recognize jokes (yes or no) or to choose the humorous excerpt from among several alternatives.

For example, suppose one of these excerpts contained a reference to a man's touching his head. How is a computer—rigidly programmed, remember—to evaluate the possible humor of this gesture? Touching one's hand to one's head may mean the person has a headache; the person is a baseball coach giving a signal to the batter; the person is trying to hide his anxiety by appearing nonchalant; the person is worried about his hairpiece slipping; or *indefinitely many* other things, depending on indefinitely many ever-changing human contexts. Humor, since it depends on so many emotional, social, and intellectual facets of human beings, is particularly immune to computer simulation.

Switching gears (and subjects), let us recall that the theory of types and subsequent constructions make the derivation of Russell's set theory paradox impossible. This method of avoiding the Cretan paradox and other natural language paradoxes is, however, a little strained and at variance with common usage. The everyday notion of truth does not come with a level number attached. Truth is truth—not $truth_2$ or $truth_{17}$. An alternative, more natural approach to the Cretan paradox is to eliminate the requirement that every sentence be either false or true. We can classify the paradoxical sentence as neither true nor false, or maybe as both true and false, or maybe even as a sort of mood signal.

It has been suggested by Zen philosophers that notions like truth and falsity, subject and object, external and internal, while essential in everyday life as well as in scientific thought, nevertheless prevent one from attaining a mystic, oceanic union with the universe. The universe simply is. Paradoxes like the Cretan paradox, since they seem to do violence to our concepts of truth and falsity, might thus be taken as a reminder of this essential is-ness of the universe —a reminder that these distinctions, in some fundamental sense, are unimportant. Be that as it may (or may not), the paradoxical sentence "This sentence is false" leads when understood in a natural way (and not dismissed as meaningless or interpreted in terms of levels) to a sort of mental oscillation. If it is true, it is false. If it is false, it is true. If it is true, it is false. . . . It is this mental oscillation that concerns me (it will be incorporated into the "catastrophe" model of humor in chap. 5) and that is the reason understanding the paradox is important to understanding humor.

W. F. Fry, Jr. (1963), and Gregory Bateson (1958) have shown that (a version of) the Cretan paradox is implicit in most humorous situations (not only in modal jokes or Russell jokes). As a psychiatrist and an anthropologist, they are sensitive to the social setting in which humor takes place. Jokes are more than lines in a book or magazine. They are a peculiar form of social interaction set off from other kinds of interaction by what Fry and Bateson call a "play frame." That a joke is being told is usually indicated by some kind of metacue. This may take the form of a different voice inflection, an arched eyebrow or a wink, the use of a dialect, a mock-serious tone, or even the explicit clause, "Have you heard the one about . . . ?"

The metacue is an integral part of the joke and qualifies whatever is being said. It says, in effect, "This whole business is unreal." This self-referential cue results in a Cretan paradox. If we take the cue seriously (as real), then, by what it means, we should not. If we do not take it seriously, then, by what it means, we are. Thus, for example, a mock-serious tone or a dialect used in telling a joke (or a more extended piece of humor) says, in effect, "This situation is unreal."

All art, in fact, has these two aspects: its content and its frame (or setting), which sets it apart from nonart and which says of itself, "This is not an everyday sort of communication. This is unreal."

In this way the joking situation itself is paradoxical, regardless of the specific joke being told. That is, to the humor generated by the punch line of the joke, the metalevel cue (gesture, inflection, dialect, etc.) adds its own paradoxical humor. This, of course, is not characteristic only of humor. The same tension is induced in the theater, for instance, where the message "This is make-believe" is put across in similar ways and involves the same kind of self-referential paradox. The (pleasant) tension induced by these metacues, let me reiterate, is over and above that generated by the punch lines of the jokes and is part of the reason transcripts of a comedian's routines are not nearly as funny as the actual performances.

I will end this chapter by briefly discussing a famous and important result in mathematical logic, Gödel's incompleteness (meta-) theorem, whose proof uses (in a nonparadoxical way) the notion of self-reference.

Basically, it says that, given *any* axiomatic system (that contains a few axioms of arithmetic), there must be statements within the system (object-level statements) that are neither provable nor disprovable from the axioms of the system; that is, there must exist statements independent of the formal system. A consequence is that there can never be any formalization of arithmetic, or for that matter of mathematics in general, that is complete (in the sense that all true statements are provable from the axioms of the system). Many mathematicians had thought that there existed a complete set of axioms of arithmetic, say, from which all true facts about whole numbers could be proved. They were wrong.

Gödel's theorem is a metatheorem, a theorem about the formal system as a whole, not a theorem within the formal system about whole numbers. Its proof is complicated, but a rough outline of it is instructive and flavorful. Briefly, what is done is to consider an axiom system containing basic arithmetic statements (among them the iterative definitions of addition and multiplication considered in the last chapter). Then certain metalevel statements about this axiom system are coded into object-level statements about numbers. This is accomplished by methodically assigning each object-level statement a unique code number. Similarly, proofs of object-level statements can also be assigned code numbers. By means of this coding, object-level statements about numbers can also be understood as expressing metalevel statements about the system or about individual object-level statements. If one is careful and clever, one can find an object-level statement about numbers that, on the metalevel, says that it itself is unprovable; that is, one can find a statement that is true if and only if it is unprovable. From the facts that the axioms are all true and that the system is consistent, it is possible to conclude that such a statement is neither provable nor disprovable from the axioms—that it is independent of them. Moreover, even if we add such a statement as a new axiom, the same proof applies to the new axiom system obtained, and we can in the same manner find a statement independent of it.

This theorem and its proof may seem far removed from the logic of humor; yet, as in the case of the Cretan and Russell paradoxes, the connection is not so tenuous. Recall the joke about the new prisoner puzzled because his fellow inmates laughed whenever one of them called out a number. He was told that the numbers were a code for certain jokes, which thus did not need to be repeated verbatim. Intrigued, the new prisoner called out "63" and was greeted by total silence. Later his cellmate explained that everything depends on how the joke is told. (I suppose this metajoke could itself be assigned a code and . . .) More generally, it is not uncommon for people to make a statement containing code words that, in effect, express the speaker's attitude (meta-level feelings) toward that statement. This phenomenon occurs in politics, literature, and advertising as well as in humor. Freud too wrote of codewords, but he used the term in the more common sense of "symbol," without the meta-level overtones. Finally, in the spirit of Gödel's theorem (and with considerable looseness), we can state the following: There is no theoretical account of humor that is not itself (on a higher level) somewhat funny and therefore incomplete.

Leaving these more esoteric matters, I will turn to some very common sorts of verbal humor—puns, spoonerisms, and reversals.

4

Humor, Grammar, and Philosophy

Reversal or permutation of the grammar of a sentence often results in humor. I will call this sometimes tiresome type of humor grammatical (or combinatorial) humor for lack of a better term. It is generally not very deep. Language being the flexible and plastic tool it is, there are indefinitely many varieties of combinatorial humor. After a discussion of some of them (spoonerisms, puns, transformations, etc.), in which I leave unanswered the question why some people groan upon hearing a pun, I will turn to a deeper sort of humor. This latter type, misunderstandings deriving from the confusion of the logic of a given statement or situation for that of another, I will refer to as philosophical humor. It is probably the type the Austrian philosopher Ludwig Wittgenstein had in mind (1953, 1958) when he remarked that a serious work in philosophy could be written that consisted entirely of jokes. One "gets" the joke if and only if one understands the relevant philosophical point.

To start, let us consider the simplest combinatorial transformation, the spoonerism. A spoonerism occurs when the sounds of two or more words in a phrase or sentence are interchanged. Examples are "I've had tea many martoonis," "Time wounds all heels," and "tons of soil" (for "sons of toil"). The anthropologist G. B. Milner (1972), using notions of the famous linguist Saussure[1] has considerably extended the notion of a spoonerism (as well as that of a pun). Thus an interchange of whole words is in a sense a generalized spoonerism. "A hangover: the wrath of grapes" and "alimony: bounty from the mutiny" are examples.

If we stretch things a bit, a relational reversal, the interchange of two objects or people standing in a certain relation to each other, may also count as a kind of generalized nonlinguistic spoonerism. Thus, for example, a greyhound dog with a bus tattooed on its side (fig. 17) is a relational reversal, as is a group of convicts in striped clothing patrolling a prison block in which all the prisoners are wearing business suits. An old joke cited by Freud is another good example. A marquis at the court of Louis XIV enters his wife's bedroom and finds her in the arms of the bishop. He sees them, then walks calmly to the window and goes through the motions of blessing the people in the street. "What are you doing?" cries the perplexed wife. "Monseigneur is performing my function," replies the marquis, "so I am performing his." Relational reversal is of course a very common gambit in comedies, humorous newspaper columns, nightclub routines, and so forth.

Reversals of this kind are often humorous because they force us to perceive in quick succession the familiar relation and an unfamiliar (and therefore incongruous) one. This notion of a relational reversal leads naturally to Gestalt psychology, which stresses the holistic nature of perception. Situations, problems, sentences are perceived as whole figures, while their background is more or less screened out. Certain well-known illustrations show, however, that perception of the main figure and its background sometimes varies depending on how one looks at the picture. Two examples are the

1. Saussure, one of the founders of modern linguistics, stressed that the meaning of a word derives in large part from the contrast between that word and other words that could take its place in the phrase or sentence in which it appears.

Fig. 17

Necker cube and the faces-urn drawing shown in figure 18. Here one way of perceiving the picture alternates easily with another.

Relational reversals (perhaps jokes in general) can be considered as a kind of Necker cube presenting us in rapid succession with a given situation and its reversal (in some sense or other). By contrast, each alternative reinforces the different meaning of the other. If the different meanings are incongruous and the emotional climate is right, humor results. Even if the emotional climate is not quite right, insight results, as in some of the mathematical examples of chapter 1.

Psychological theories that, like Gestalt, stress the cognitive, intellectual aspects of humor contribute more to an un-

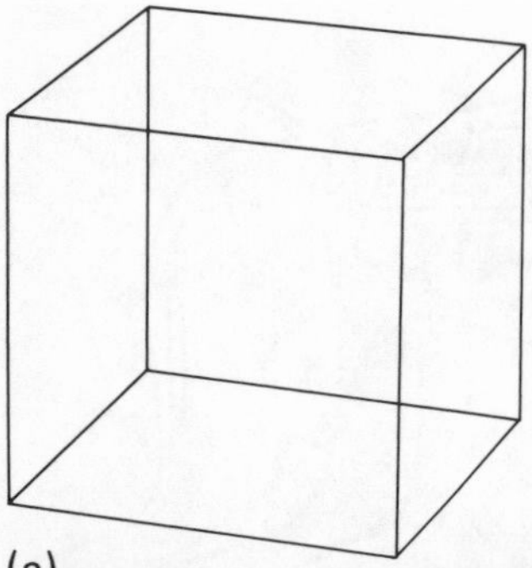

(a)

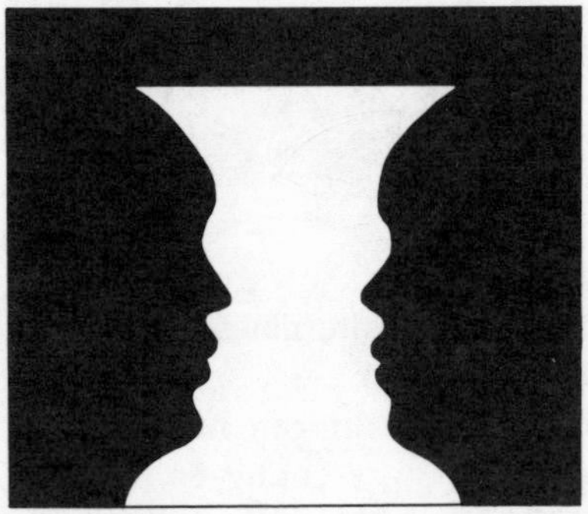

(b)

Fig. 18

derstanding of humor, it seems to me, than do behaviorist theories or plain blind empiricism. The reason is simple: humor usually has a cognitive, intellectual component. Moreover, even the affective or emotional component seems more amenable to even a Freudian or some sort of "humanistic" analysis than to a behavioristic one.

The psychologists Suls (1972), McGhee (1978), and Shultz (1976), among others, have made studies that tend to confirm the important role cognitive development plays in the resolution of incongruities. Children do not appreciate certain types of jokes until they have mastered the relevant intellectual machinery. Similarly, adults do not laugh until they have resolved a joke's incongruity, which resolution depends on several factors—complexity, arousal level, focus, and so forth. The work of the psychologist Piaget, though not directly concerned with humor, also emphasizes cognition and the pleasure of mastery. Behaviorism,[2] on the other hand, has spawned much research attempting to find statistical correlations between variables, usually operationally defined in some more or less silly way. It ignores intentions, context, values, and so forth. Even when this succeeds, the result obtained has little explanatory value since it is not imbedded in any theoretical framework.

Probably one of the most common varieties of grammatical humor is the pun. Words (and phrases) are usually classified into clusters of words that "belong together" for one reason or another. A pun provides a link between two or more distinct clusters (universes of discourse) by means of a word or phrase that has a different meaning in each. Usually this is accomplished by using homonyms.

Consider the following two puns: "Colds can be positive or negative. Sometimes the ayes have it, sometimes the noes."

2. Formalism, a philosophy of mathematics that claims roughly that mathematics can be reduced to operating on and manipulating meaningless marks on a piece of paper, is in some ways analogous in its reductionist emphasis to behaviorist philosophies in psychology.

Interviewer: "Do you consider clubs appropriate for small children?" W. C. Fields: "Only when kindness fails." In the first, "ayes" and "noes" provide a link between the word cluster having to do with parliamentary rules and that relating to cold symptoms. In the second, which is funnier (probably because it is more aggressive), "clubs" can refer either to Little League, Girl Scouts, and other social organizations or to blunt instruments, beating, and so on.

Like the relational reversal, a pun forces one to perceive in quick succession two incongruous and unrelated sets of ideas. The suddenness is, as in much of humor, very important. Explaining a pun, or humor in general, often kills it. This will be partially accounted for in terms of my model for jokes and humor in the next chapter.

A convenient way to conceive of puns is in terms of the intersection of two sets. A pun is a word or phrase that belongs to two or more distinct universes of discourse and thus brings *both* to mind. The humor, if there is any, results from the inappropriate and incongruous sets of associated ideas jarring each other. Thus the W. C. Fields pun related above can be pictured as in figure 19, where the word *clubs* can be

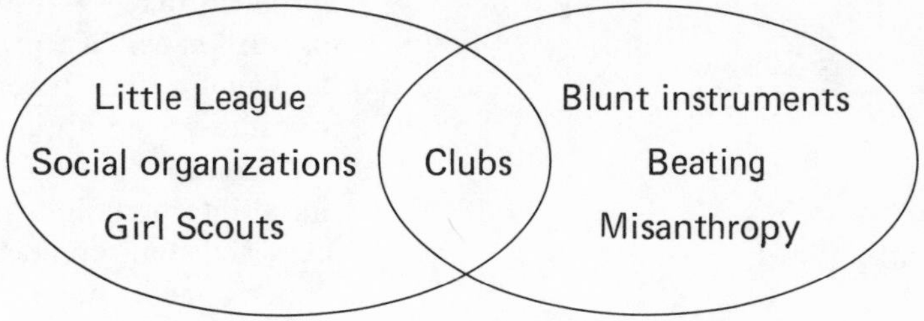

Fig. 19

seen as forcing one to juxtapose the two unrelated sets of ideas. The energy flow, so to speak, is from left to right in the diagram, as *clubs* serves as a slide down which the laughter falls. In chapter 5 this metaphor will be explained more precisely.

Every word is a member of indefinitely many different clusters, since it can be classified according to indefinitely many different criteria. Thus, determining whether a sequence of words constitutes a pun (in particular a good pun) depends, as in deeper forms of humor, on meaning (context, values, intentions, etc.). Multiple (intersection of more than two word clusters) and layered (conflation of levels) puns add more complexity to this puny form of humor. Also, if certain word clusters have a large overlap (intersection), a whole series of related puns may be developed. Culinary and sexual clusters, for example, provide fertile ground for such a series. Last, the notion of a pun, like that of a spoonerism, can be generalized to nonlinguistic categories. An analysis in terms of intersections also applies to this generalization.

The contrast in meaning between a figurative interpretation of a statement and a literal one is the source of much verbal humor. Groucho Marx's quip, "When I came to this country I hadn't a nickel in my pocket. . . . Now I have a nickel in my pocket," is a well-known and typical example. Roughly, the pattern is to make a statement and then repeat it in some way or other, the second time stressing the alternate (literal or figurative) interpretation. Mathematicians, among others, are much taken with this practice. The injunction often seen on trash cans, "Keep Litter in Its Place" has always particularly amused me. If something is litter, its place by definition, it seems, *is* the ground.

A contrast more general than that between figurative and literal interpretations is that between a phrase, statement, or story and some combinatorial permutation of it. Here a statement is made, then rearranged and reiterated, the related form and different meaning being stressed the second time. This technique (known as chiasmus) was also studied by Milner (1972) in a Saussurian framework; it is not of course limited to humor. "Ask not what your country can do for you. Ask rather what you can do for your country." "If you must behave like a lunatic in school, you'll have to behave like a student in the asylum." These are examples of chiasmus. The permutations and parallel constructions of these examples function like a relational reversal or a pun: they bring to mind different universes of discourse almost simultaneously. Furthermore, their concise deftness sometimes gives added pleasure.

Good examples of chiasmus are common in books of epigrams and quotations; bad ones are common in the daily conversation of certain wearying people. Classifying the structure of every possible chiasmus is an unrewarding if not impossible job, since almost any statement can be rearranged to yield an intelligible chiasmus (except maybe this one). In certain situations the first line of a chiasmus, if it is very familiar, is suppressed but tacitly understood. Often the "second" line is the same as the unstated first line but with a different intonation and emphasis indicating scorn, skepticism, or irony.

Nonsense sounds are still another common form of verbal humor. Here the contrast between the seemingly meaningful and suggestive sounds and their complete lack of meaning is very pouse to the purk of the tumor. Lewis Carroll's nonsense verse "Twas brillig, and the slithy toves / Did gyre and gimble in the wabe" is a good example. Almost any passage from James Joyce's *Finnegans Wake* provides further examples of nonsense wounds (biting humor), puns, and malapropisms. Nonsense sounds (or wounds) with no hint of meaning or form, however, are usually just annoying unless they are particularly euphonious. Linocera.

There is an interesting algorithm or recipe (one of many devised by a group of French writers) for generating nonsense that is often meaningful-sounding and sometimes mildly humorous. Take a famous passage and substitute for every other noun the eleventh noun following it in your dictionary. "In the being God created the heavyweight and the earth. And the earwax was without form, and void; and darning was upon the face. . . ." Similarly, consistently substituting one appropriately inappropriate word for another in a paragraph or a story is sometimes humorous.

I will conclude this survey of grammatical humor with a brief sketch of Noam Chomsky's theory of transformational grammar. Although this theory is not concerned with humor, it does provide us with some general ideas useful in the study of (grammatical) humor. A basic notion is the distinction between the surface structure of a sentence and its deep structure. The surface structure is the structure we hear spoken or see written. It is not sufficient to account for all the syntactic (grammatical) or semantic (meaning) features of the sentence. A deep structure that does account for all these features is hypothesized to exist. In rough terms, it expresses the logic of the statement stripped of the confusing grammar of the surface structure. The deep structure is changed into the surface structure by means of transformational rules (hence the term transformational grammar). These quasi-innate rules presumably are used in a more or less unconscious way by speakers, listeners, and readers of the language to unscramble the meaning of sentences.

For example, the two sentences "John is eager to leave" and "John is difficult to leave" have quite different deep structures, but similar surface structures. *John* is the grammatical (surface structure) subject of both but is the logical (deep structure) subject only in the former. In the latter, *John* is the logical object. (What the deep structures are, and what the transformations are that change them step by step into the two sentences above, is better left to a course in transformational grammar.) Another example of sentences with similar surface structures but dissimilar deep structures is the pair "The sugar is slow to dissolve" and "The sugar is easy to dissolve."

If more than one deep structure can be associated with a statement having a given surface structure, the statement is ambiguous. Thus, "Mortimer knows a kinder person than

Waldo" may be short for either of the following sentences: "Mortimer knows a kinder person than Waldo knows" or "Mortimer knows a kinder person than Waldo is." These two sentences clearly have different deep structures, yet both these deep structures can be changed into "Mortimer knows a kinder person than Waldo" by transformational rules. Hence the original statement is ambiguous. Similarly, "All that glitters is not gold" may mean "Not everything that glitters is gold," or it may mean "Nothing that glitters is gold." "The shooting of the hunters was dreadful" is a final example of an ambiguous statement having very different deep structures.

Of what relevance is this to humor? It is yet another linguistic device (and happily the last I will consider) for calling to mind two different interpretations in quick succession. Humor often results. An old example is the story of the cannibal returning home one evening, asking if he is late for dinner, and being told, "Yes, everybody's eaten." The reply, it is clear, has two deep structures, one in which "everybody" is the subject of the verb *eat*, the other in which it is the object. Another example concerns an idiot driving to Chicago. He comes to a sign, CHICAGO LEFT, swears to himself, then turns around and heads back home. The sign is ambiguous in the same way: two different deep structures are associated with it.

A related sort of ambiguity occurs when the words of the surface structure sentence can be grouped in more than one way, thus yielding more than one associated deep structure. For example, someone asks a farmer how long cows should be milked, and the farmer replies, "the same as short ones, of course." Here "(how long) cows" was interpreted by the farmer as "how (long cows)." Or consider the following exchange. Wife (or husband): "Won't you give up smoking for me?" Spouse: "Why do you think I'm smoking for you?" Here "(give up smoking) (for me)" is understood as "(give up) (smoking for me)."

I have limited myself largely to explicating the logic or grammar of verbal humor for two reasons: this is a book and not a theater, television set, or gallery; and verbal jokes lend themselves more easily to a mathematical or, in the case of grammatical jokes, a quasi-mathematical treatment than do nonverbal jokes. Nevertheless, some of what I have said can be widened in scope if the definition of logic and grammar is extended to include situations, physical movements, musical or visual arrangements, and so forth.

Thus Picasso's *Bull*, in which the seat and handlebars of a bicycle suggest the head of a bull, is a visual pun linking two different sets of images. Magritte's paintings *Not to Be Reproduced*, and *Evening Falls*, with their strange mirrors and windows, are reminiscent of the notions of self-reference and levels. Slapstick and physical humor have a logic of their own (repetition, exaggeration, inappropriate dress, etc.). The dignified movements of Charlie Chaplin clash humorously with his appearance as a powerless little man. The relational reversals cited earlier in the chapter (prisoners in place of guards, marquis in place of bishop) are further examples of nonverbal humor. Similarly, substituting a piccolo for a bass viol in a symphony or a bass for a soprano in an opera, or mixing incongruous themes in any "heavy" piece is likely to be humorous. Compare P. D. Q. Bach.

In each of these cases a "logic" or "grammar" not of statements but of situations, movements, visual images, or musical forms is tacitly understood. The people who study and articulate such "logics" are not mathematicians or logicians, of course, but novelists, critics, artists, and musicians. Nevertheless, this division of labor should not, I think, be taken too strictly. There is no good reason mathematicians should refrain from applying mathematical notions to painting, music, and literature or, conversely, refrain from using ideas or techniques from these disciplines to suggest new mathematical structures and operations. The same holds true for artists, musicians, writers, and critics. (Many ideas in Borges, for example, seem to suggest strange new mathematical structures.)

Moving from artists, critics, and musicians to philosophers, recall Ludwig Wittgenstein's remark that a serious work in philosophy could be written that consisted entirely of jokes. He meant, of course, that "getting" certain jokes is possible if, and only if, one understands the relevant philosophical point. Let us now examine some of this "philosophical humor."

George Pitcher (1966) has demonstrated some very interesting similarities between the philosophical writings of Wittgenstein himself and the work of Lewis Carroll. Both were concerned with nonsense, logical confusion, and language, although, as Pitcher notes, Wittgenstein was tortured by these things whereas Carroll was (at least in his writings) delighted by them. Pitcher cites many passages in *Alice in Wonderland* and *Through the Looking Glass* as illustrating the type of joke Wittgenstein probably had in mind when he made the comment referred to above.

The following excerpts are representative of the many in Lewis Carroll that concern topics that Wittgenstein wrote about and that demonstrate a purposeful confusion of the logic of the situation.

> 1. She [Alice] ate a little bit, and said anxiously to herself, "Which way? Which way?" holding her hand on the top of her head to feel which way it was growing, and she was quite surprised to find that she remained the same size. [*Alice in Wonderland*, p. 10]

> 2. "That is not said right," said the Caterpillar.
>
> "Not *quite* right. I'm afraid," said Alice timidly; "some of the words have got altered."
>
> "It is wrong from beginning to end," said the Caterpillar decidedly, and there was silence for some minutes. [*Alice in Wonderland*, p. 47]

3. "Then you should say what you mean," the March Hare went on.

"I do," Alice hastily replied; "at least—at least I mean what I say—that's the same thing, you know."

"Not the same thing a bit!" said the Hatter. "Why, you might just as well say that 'I see what I eat' is the same thing as 'I eat what I see'!" [*Alice in Wonderland*, pp. 68–69]

4. "Would you—be good enough," Alice panted out, after running a little further, "to stop a minute just to get —one's breath again?"

"I'm *good* enough," the King said, "only I'm not strong enough. You see, a minute goes by so fearfully quick. You might as well try to stop a Bandersnatch!" [*Through the Looking Glass*, pp. 242–43]

5. "It's very good jam," said the Queen.

"Well, I don't want any *to-day*, at any rate."

"You couldn't have it if you *did* want it," the Queen said. "The rule is jam to-morrow and jam yesterday—but never jam to-day."

"It *must* come sometimes to 'jam to-day,' " Alice objected.

"No, it can't," said the Queen. "It's jam every *other* day; to-day isn't any *other* day, you know."

"I don't understand you," said Alice. "It's dreadfully confusing." [*Through the Looking Glass*, p. 206]

What do these examples have in common? As noted, they all betray some confusion about the logic of certain notions. One does not lay one's hand on top of one's head to see if one is growing taller or shorter (unless only one's neck is growing). One cannot recite a poem incorrectly "from beginning to end," since then one cannot be said to be even reciting that poem. (Wittgenstein was very concerned with criteria for establishing identity and similarity.) In the third quotation the Mad Hatter is presupposing the total independence of meaning and saying, an assumption that Wittgenstein shows leads to much misunderstanding. The fourth passage confuses the grammar of the word *time* with that of a word like *train*, and the fifth illustrates that the word *today*, despite some similarities, does not function as a date. Both these latter points were also discussed by Wittgenstein.

Wittgenstein explains that "When words in our ordinary language have prima facie analogous grammars we are inclined to try to interpret them analogously; i.e. we try to make the analogy hold throughout." In this way we "misunderstand . . . the grammar of our expressions." These linguistic misunderstandings can be, as I have mentioned, either sources of delight or sources of torture depending on one's personality, mood, or intentions. Wittgenstein was concerned (tortured even) by the fact that a person does not talk about having a pain in his shoe even though he may have a pain in his foot and his foot is in his shoe. Carroll, had he thought of it, probably would have written of shoes so full of pain that they had to be hospitalized.

Open any book on analytic philosophy and you will find clarifying distinctions that, if utilized differently, could be the source of humor. The following pairs of phrases serve as examples of what I mean. "Going on to infinity" versus "going on to Milwaukee"; "honesty compels me" versus "my mother compels me"; "the present king of France is hairy" versus "the present president of the United States is hairy"; "an alleged murderer" versus "a vicious murderer"; "Have you stopped beating your wife?" versus "Have you voted for Kosnowski yet?" "before the world began" versus "before the game began." The first phrase in each case shares the same grammar as the second phrase, yet the logic (in a broad sense) of the two is quite different.

In fact, much of Wittgenstein and modern analytic philosophy in general has been concerned with unmisunderstanding (getting clear about) the logic and (surface) grammar of problematic terms (e.g., time, mind, rule, action, pain, reference) as well as with explicating and clarifying phrases such as the ones in the previous paragraph. Analytic philosophy can in a sense even be called linguistic therapy, and philosophers like Wittgenstein, Ryle, and Austin have devoted much effort and analysis to curing some of these linguistic diseases. Pitcher comments that Alice is a victim of the characters in her mad world of nonsense just as the philosopher is the victim of the nonsense he unknowingly utters. Wittgenstein (1956) writes, "The philosopher is the man who has to cure himself of many sicknesses of the understanding before he can arrive at the notions of a sound human understanding. If in the midst of life we are in death, so in sanity we are surrounded by madness." In humor the anxiety induced by these misunderstandings as well as by more traditional philosophical concerns (God, death, choice) finds its release in laughter. (Compare Woody Allen and Kierkegaard, say, or the "humor" of Samuel Beckett.)

Speaking of Woody Allen, an excerpt from his analysis of ink blots is totally out of place here and will therefore be inserted.

> The first ink blots, it was learned, were crude, constructed to eleven feet in diameter and fooled nobody.
>
> However, with the discovery of smaller size by a Swiss physicist, who proved that an object of a particular size could be reduced in size simply by "making it smaller," the fake ink blot came into its own.
>
> It remained in its own until 1934, when Franklin Delano Roosevelt removed it from its own and placed it in someone else's. [Allen 1972]

At the risk of stretching the connection to the logic of humor, I will end this chapter by discussing two fascinating paradoxes from the philosophy of science that are not exactly funny but at least bring a smile to the cerebrum. They are tangentially relevant to chapter 6 as well. Both concern scientific induction, the establishing of empirical statements as true or at least probable. (The Scottish philosopher Hume in the eighteenth century noted that an inductive justification of scientific induction is circular. The two paradoxes quoted here are independent of this observation.)

Carl Hempel's "raven" paradox, so called because it is usually illustrated with ravens, can be easily stated. Suppose one wants to confirm the statement that all ravens are black. One goes out, looks for ravens, and checks to see if they are black. We believe that if we observe enough instances of black ravens, we will have confirmed (not necessarily conclusively verified) the statement "All ravens are black." But by elementary logic "All ravens are black" is logically equivalent to "All nonblack objects are nonravens." Since the statements are equivalent, any observation that confirms one confirms the other. But pink flamingos, orange shirts, and chartreuse lampshades are all instances of nonblack objects and thus tend to confirm the statement "All nonblack objects are nonravens." Thus they must also confirm "All ravens are black." Hence we arrive at the curious position of having pink flamingos, orange shirts, and chartreuse lampshades confirming the statement that all ravens are black! (The emotional climate is not right for humor. This is just odd, not funny.)

What is the problem? Well, it still is not clear to people. Two quick points should be made, however. One is that merely amassing instances of a statement is not enough to confirm it. The second is that nonravens and nonblack objects are much more numerous than ravens and black objects. Perhaps we could understand pink flamingos, orange shirts, and chartreuse lampshades as confirming, but only *very* slightly, the two equivalent statements above—not as much, in fact, as a black raven would.

The second paradox, due to Nelson Goodman (1965), is called the grue-bleen paradox and concerns the odd color terms *grue* and *bleen*. An arbitrary future date is selected, say 1 January 2001. An object is defined to be grue if it is green and the time is before 2001 or if it is blue and the time is after 1 January 2001. Something is bleen, on the other hand, if it is blue and the time is before 2001 or if it is green and the time is after 1 January 2001. Now consider the color of emeralds. All emeralds examined up to now (1980) have been green. We therefore feel confident that *all* emeralds are green. But all emeralds so far examined are also grue. It seems that we should be just as confident that all emeralds are grue (and hence blue beginning in 2001). Are we?

A natural objection is that these color words grue and bleen are very odd, being defined in terms of the year 2001. But were there a people who speak the grue-bleen language, they could make the same charge against us. "Green" is an arbitrary color word, being defined as grue before 2001 and bleen afterward. Blue is just as odd, being bleen before 2001 and grue from then on.

What exactly is wrong with the terms grue and bleen has not yet been convincingly established by philosophers. This is not the worst of our problems, however, since, as Woody Allen points out, "Not only is there no God, but try getting a plumber on weekends."

5

A Catastrophe Theory Model of Jokes and Humor

Running through this account of the logic of humor has been the idea of an abrupt switch or reversal of interpretation resulting in the sudden perception of some situation, statement, or person in a different and incongruous way. This interpretation switch may be accompanied by the overcoming of a mild fear or anxiety, as when one realizes that what seemed threatening is really not so, or perhaps as when one solves a riddle. Often the release of hostile feelings accompanies the switch, as when one makes an aggressive or a sexually offensive witticism. At other times the interpretation reversal signals the expression of a playful approach to a situation. At still other times, the achieving of self-satisfaction is a concomitant of the reversal, as in the "sudden glory" resulting from someone else's (mild) misfortune. In all these cases we have a sudden interpretation switch bringing about a release of emotional energy, the release usually taking the form of laughter.

A very interesting topological theory recently discovered by the French mathematician René Thom (1975) concerns itself with the description and classification of such discontinuities (jumps, switches, reversals). This theory, known as catastrophe theory, provides a sort of mathematical metaphor for the structure of humor and will help us in particular to visualize that structure more clearly. Fortunately, to appreciate some of its applications it is not necessary to understand the proof of its main theorem. Nevertheless, the preliminaries require a certain amount of exposition.

The first mathematical notion that is needed is that of a three-dimensional coordinate system (see fig. 20). A point

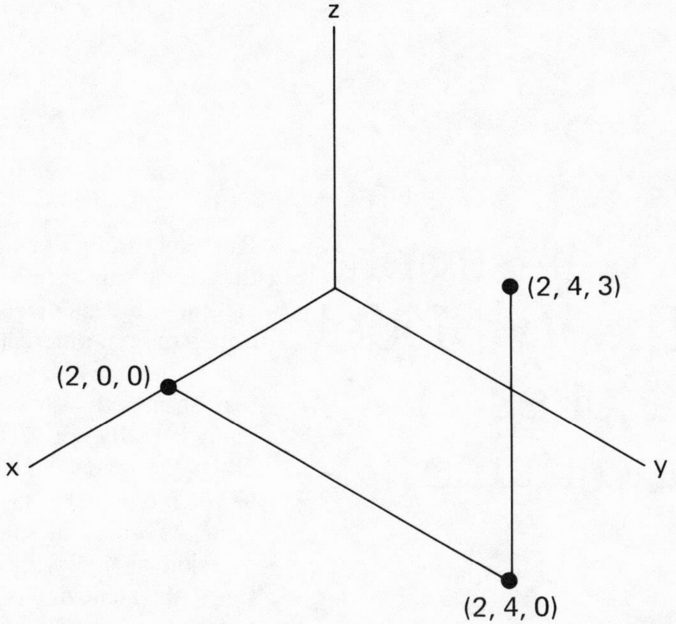

Fig. 20

in this three-dimensional space is located by specifying its x, y, and z coordinates—the distances one must measure in the x, y, and z directions to locate the point. Thus (2, 4, 3) are the coordinates of a point located by moving from the origin (the intersection of the perpendicular axes) 2 units in the x direction, 4 units in the y direction, and then 3 units up in the z direction. It should be clear that every point in space has coordinates of this form and that every set of three numbers corresponds to some point in space.

One of the most common types of model arising from catastrophe theory can best be understood by an example adapted from one by E. C. Zeeman (1976). Zeeman, an English mathematician, has discovered many ingenious applications of these models. This one concerns aggressive behavior in animals (dogs in particular), which has been shown to depend (largely) on two factors: fear and rage. Fear by itself induces the dog to retreat, whereas rage by itself induces the dog to attack. The absence of fear and rage, of course, results in neutral behavior. A very interesting phenomenon is that high levels of both fear and rage *together* rarely result in neutral behavior, but rather lead to either attack or retreat depending on how the fear and rage were built up (see fig. 21). Furthermore, and this is *very important*, increasing the fear of a growling or attacking dog slightly often abruptly sends the animal fleeing. Likewise, increasing the range of an avoiding or retreating dog slightly often causes it to suddenly attack.

Let us now consider a three-dimensional space with the x, y, and z coordinates corresponding to numerical measures of fear, rage, and behavior, respectively. The fear an animal feels can be roughly quantified, a low number indicating little fear, a higher number more fear. In dogs, the extent to which the ears are flattened back is a rough measure of fear. The rage a dog feels can be assigned a numerical value in the same way, according to how wide the mouth is opened. The behavior ranges gradually from flight through avoidance, neutrality, and growling to attack. The behavior receives a higher numerical value as one progresses through the sequence from flight to attack.

Now, given any pair of values (x,y) for fear and rage, there is at least one likely type of behavior, z. Let's indicate this by $z = f(x,y)$; z is a (possibly two-valued) function of x and y. In general, if there is much fear (large x) and little rage (small y), there is only one likely value for the behavior, a small value for z indicating flight (or at least avoidance).

Similarly, if y is large and x is small, there is only one likely value for z, a large number indicating attack (or at least growling). If both x and y are small, $z = f(x,y)$ is a "medium" value indicating neutrality. *But*, if x and y are both large (much fear and much rage), there are *two* likely values for $z = f(x,y)$—one large and one small, indicating either attack or flight (see fig. 22).

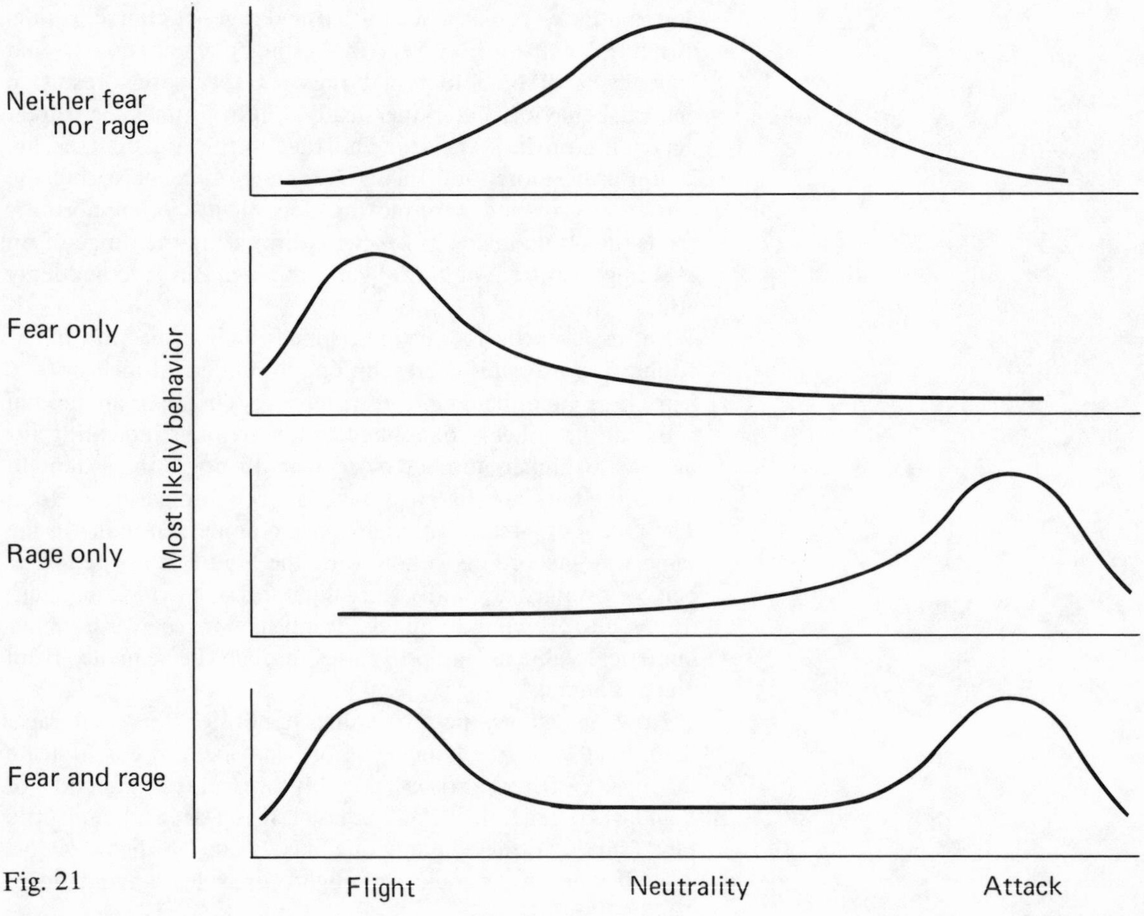

Fig. 21

Fig. 22

For each pair of numbers, x and y, we plot the point (x, y, z) where $z = f(x,y)$ is (one of) the most likely value(s) corresponding to the given pair x and y. The resulting graph consists of all these points, each of which represents the (a) likely outcome associated with a given pair of numbers, x and y. The graph is a surface in the three-dimensional space. It is now a consequence of Thom's main theorem that this surface *must* have a definite, very distinct shape (shown in fig. 23). The theorem states that any behavior that depends on two factors, is discontinuous, and satisfies two mild conditions[1] *must* give rise to this shape. Since we are interested only in the *qualitative* shape of the graph, details of how to accurately measure rage, fear, and behavior are unimportant.

There is a double layer in the middle of the surface, progressively narrowing to a point. The double layer is, as we shall see, what gives the surface its most distinctive properties. The region over which there are two layers (likely behaviors) is indicated in the x–y plane by a cusp-shaped curve. It is the region where fear and rage are both high. This model is thus called the cusp catastrophe.

To get a feel for this model, let us examine some of its properties. In the example developed so far, we know empirically that if an attacking dog is made slightly more fearful or a little less enraged, its behavior may undergo a catastrophic change (in the sense of being sudden, abrupt, and of relatively large magnitude). What happens pictorially in terms of the model is that the dog's behavior "falls off" the top layer to the bottom layer, precipitating the sudden change from attack to flight (fig. 24). Likewise, a fleeing yet enraged dog may, upon being goaded a little more, suddenly

1. Required is (1) that the behavior z at any point (x,y) be the (a) most likely outcome associated with (x,y) as in our example, and (2) that the function expressing the likelihood that an arbitrary behavior z will occur at arbitrary point (x,y) be a "smooth" function, thus enabling one to use tools from the calculus.

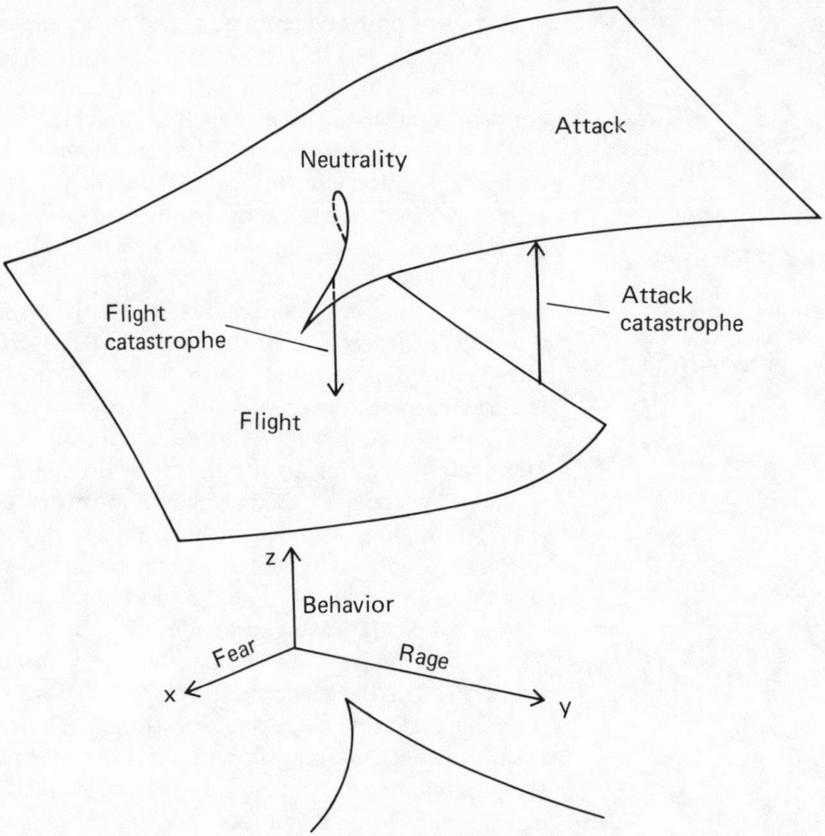

Fig. 23

turn and attack. In terms of the model, his behavior "jumps" to the top layer.

In this situation the cusp catastrophe model fits the facts. I should reiterate that Thom's theorem states, among other things, that *any* behavior or quantity depending on two factors, having a discontinuity, and satisfying certain mild general conditions (as explained earlier) must, when graphed, give rise to the same general shape. It is from this uniqueness condition that the theorem derives its power in applications.

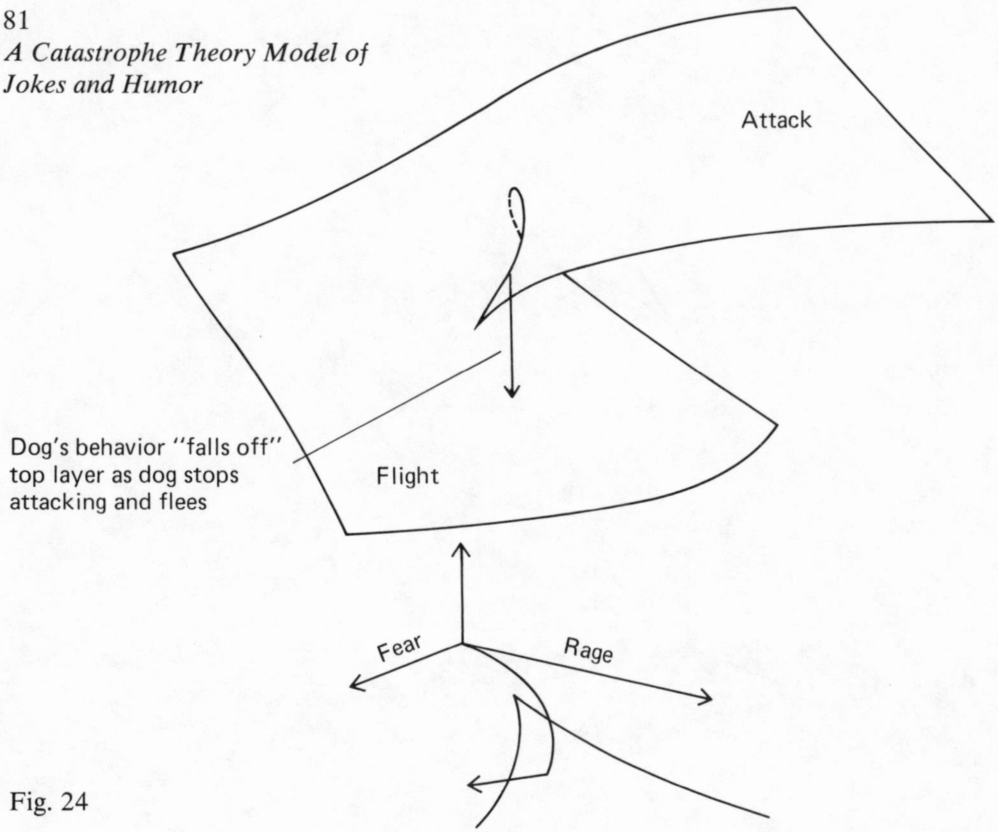

Fig. 24

Returning to the example, note that if both rage and fear are high, the behavior exhibited depends on the way the fear and rage were built up. Thus, if at first a little fear was induced and then both rage and fear were increased to certain levels, say x and y, the resulting behavior might be flight. But if a little rage was first induced, and then both rage and fear were increased to the *same* values x and y, the resulting behavior might well be attack. This is illustrated in figure 25. This property, called divergence, makes the cusp catastrophe

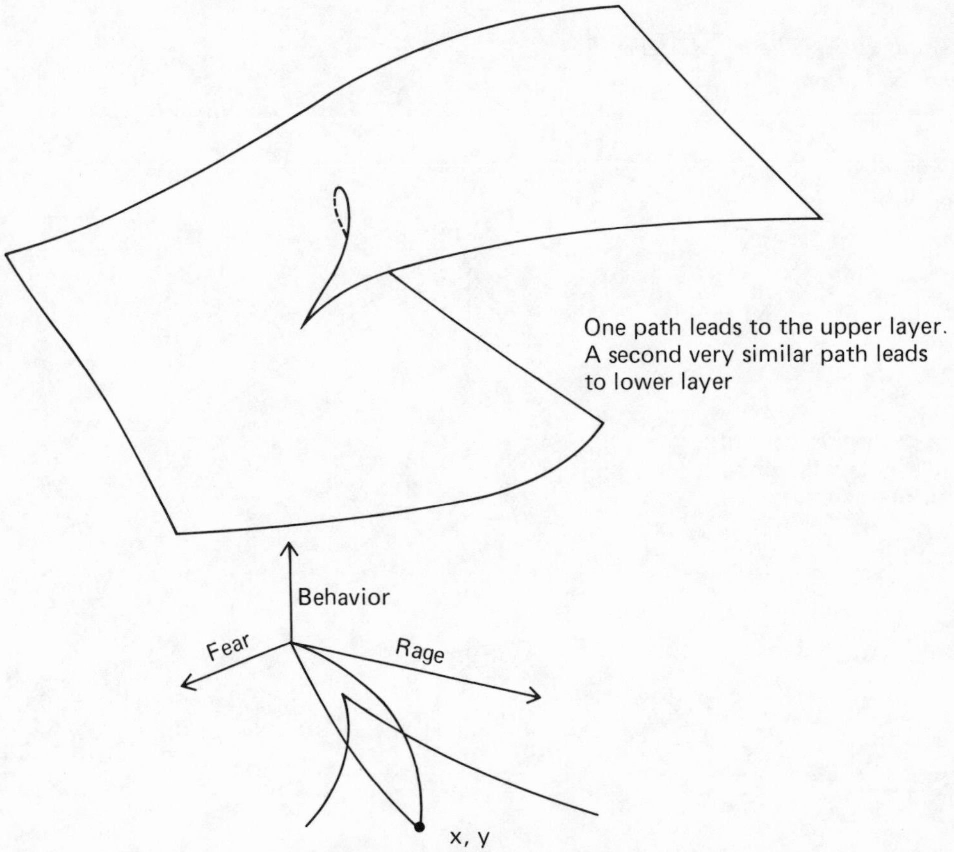
One path leads to the upper layer.
A second very similar path leads
to lower layer
Behavior
Fear
Rage
x, y

Fig. 25

particularly useful in the social and biological sciences, where behaviors, responses, attitudes, and so forth, in addition to being subject to abrupt and discontinuous changes, sometimes vary greatly despite almost identical "causes."

Another important consequence of the shape of the surface is that while a small change in one or both of the x and y coordinates may bring about a large and abrupt change in z, reversing this small change will not reverse the large change in z. Thus, in our particular example, if a little more goading

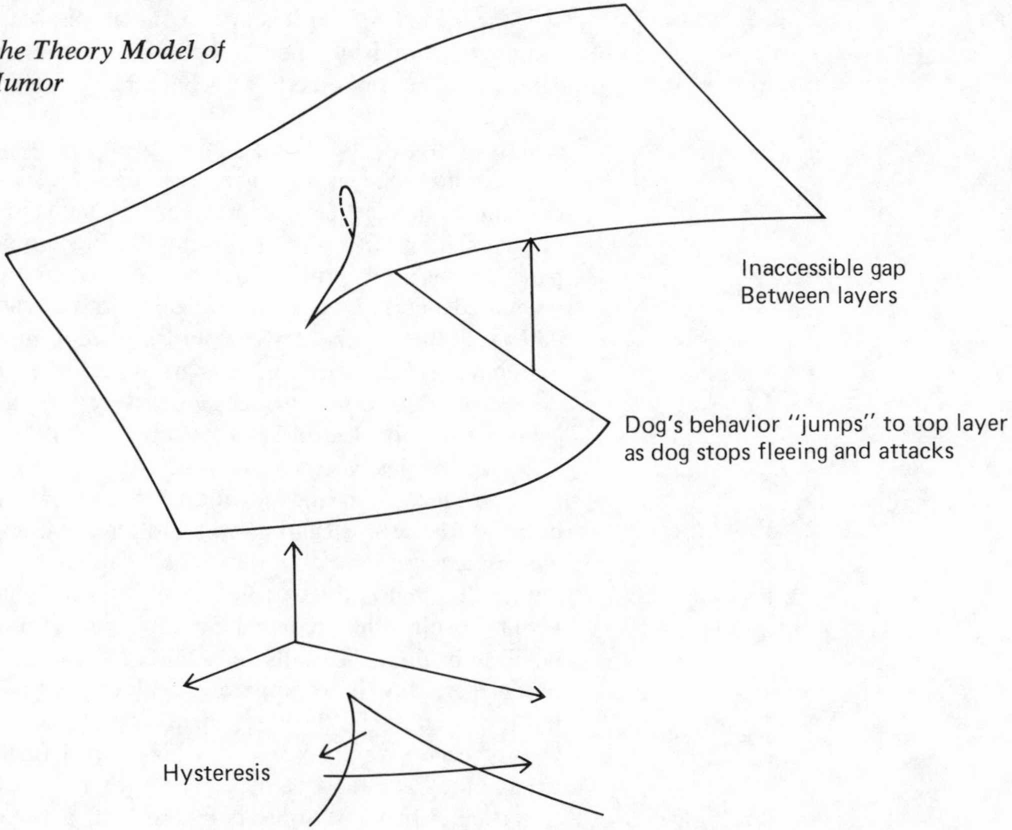

Fig. 26

finally precipitates an attack, decreasing the dog's rage by a bit (or increasing its fear a little) will *not* end the attack. To do this will require a relatively much larger change in rage and fear. This phenomenon, called hysteresis, is illustrated in figure 26. Finally, note that there is an inaccessible gap in the behaviors possible when fear and rage are both high, indicating that a neutral behavior is very unlikely.

All these properties—catastrophic jumps between two layers, divergence, hysteresis, and an inaccessible gap—are consequences of the general shape of the graph; and the general shape is dictated by Thom's theorem, which stipulates, as I have said, that any quantity (behavior) that depends on two factors, is discontinuous, and satisfies certain mild general conditions must give rise to this shape. Zeeman (1976) and others have studied many such quantities. The price index may be thought to depend (largely) on excess demand and degree of speculation. In some circumstances one's mood may depend on one's anxiety and frustration. National defense policy may depend on the factors of territorial threat and cost of defense. Whether the catastrophes are stock market crashes or recoveries, cathartic releases of anger or anxiety attacks, decisions to go to war or decisions to stop fighting, the cusp catastrophe model is often applicable and its various properties are suggestive (but not predictive).

I introduced this model, apart from its intrinsic interest, because I believe it (and considerably more complicated and convoluted versions of it) can be adapted to the study of humor. To that end let us first consider ambiguities. An ambiguity results when a statement or story has more than one possible meaning. Usually only one of these meanings is apparent (or, if both are apparent, only one is understood in a particular context). The statement or story in which the ambiguity occurs can, however, be developed further so as to change the likelihood of the ambiguity's being interpreted in a particular way. At some point, in fact, a person suddenly (discontinuously) changes his understanding (gestalt) of the ambiguous story, and there is an abrupt interpretation switch.

This should suggest that the notion of ambiguity can be modeled by the cusp catastrophe. As an ambiguous story develops, elements are added that contribute to both possible interpretations of it. These can often be roughly quantified so as to give a measure (x,y) of the development of each of the two possible meanings of the unfolding story. (There is no unique way of doing this, but *any* way of doing it will yield the same qualitative picture, which is all we are interested in anyway.) The behavior associated with any such unfolding story is the interpretation given it (by some person or group of persons) at the given point in the story. It, too, can often be assigned a rough numerical measure, z—high values for interpreting the story in terms of the first meaning, low values for interpreting it in terms of the second meaning. For each pair of coordinates (x,y) there is at least one likely behavior (interpretation) $z = f(x,y)$. If the set of points (x, y,z) such that $z = f(x,y)$ is graphed, by Thom's theorem (the general conditions are quite plausible here) and the assumptions above we obtain the characteristic surface associated with the cusp catastrophe.

What, finally, does this say about jokes and humor? A joke, as we have seen, depends on the perception of incongruity in a given situation or its description. A joke can thus be considered a kind of structured ambiguity, the punch line precipitating the catastrophe of switching interpretations. It adds sufficient information to make it suddenly clear that the second (usually hidden) meaning is the intended one (see fig. 27).

Consider the "penguin" joke of chapter 2 as an example. Here the developing first meaning is that of a woman and a certain life-style (pictorially, gradually ascending the upper layer over the ambiguous region). The punch line, "the computer sent him a penguin," reveals the hidden second meaning and brings about the catastrophe of switching interpretations (pictorially, dropping from the upper to the lower layer of the graph). Likewise, in the W. C. Fields pun in the previous chapter, the first line. "Do you consider clubs appropriate for small children?" suggests concern for the socialization of children; the most likely behavior (interpretation) is on the upper layer of the graph over the ambiguous region. The punch line, "Only when kindness fails," reveals the hidden other meaning, and there is a catastrophic fall from the upper to the lower layer of the graph.

The properties characteristic of phenomena describable by the cusp catastrophe are informative. The catastrophic interpretation switch between layers, I have mentioned. Di-

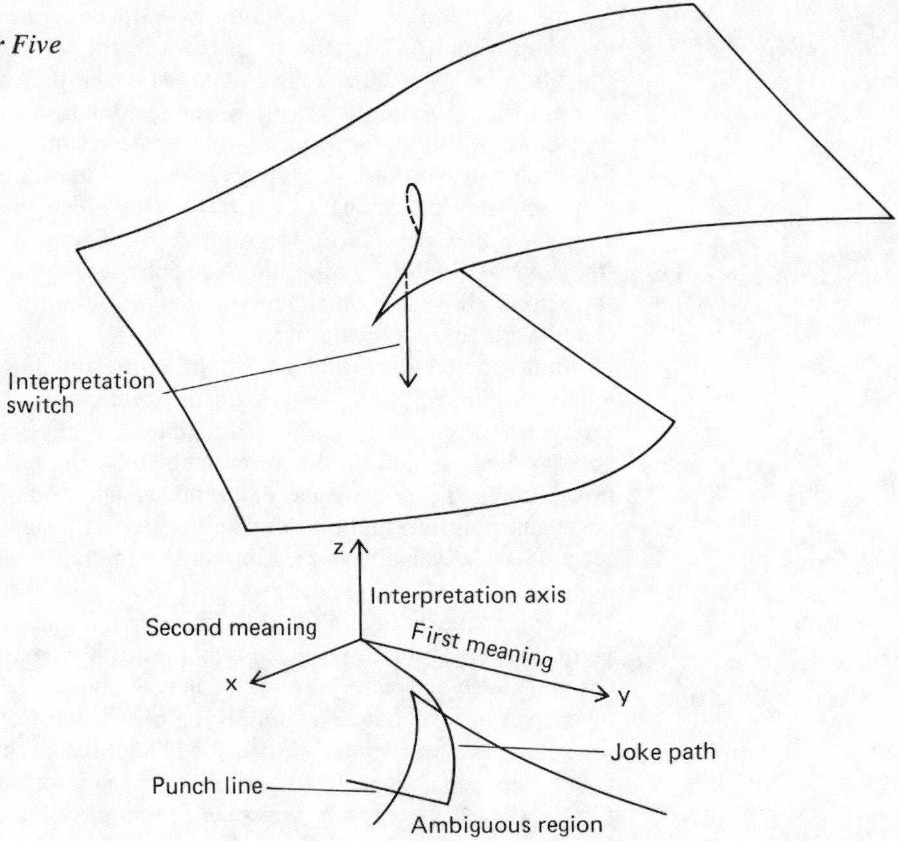

Fig. 27

vergence "explains" pictorially why minor deviations in the beginning of a story often result in its lack of humor. It "falls flat," "dies," "never gets off the ground" (see fig. 28). These phrases can be understood literally to mean that the story develops in such a way that its interpretation crosses the wrong side of the cusp and remains on the lower layer of the surface. The "buildup" of the joke fails.

The property of hysteresis demonstrates that if the alternate interpretation is given away too soon, say by relating details in the wrong order, then a very large effort is needed to reestablish the first interpretation so that the joke can proceed. Usually this is difficult or impossible. This also partially accounts for the unfunniness of jokes that must be explained, since if both interpretations are carefully explicated there is little chance for the joke to develop one interpretation and then switch suddenly to the other (fig. 29).

The shape of the surface also partially "explains" the importance of timing in the presentation of jokes. A comedian must sense how his audience has interpreted what he has already said—where the audience is located, graphically speaking, on the surface. If it is ahead of him, the alternate interpretation will become obvious too soon and the joke will lose its zing. If he is ahead of the audience, the punch line will not bring about the interpretation switch, and he will wonder why he didn't become a doctor.

The inaccessible gap in the surface illustrates the fact that only one or the other interpretation can be made at a time. Rapid alternation between the two is possible, but, as in the case of the Necker cube and other ambiguous pictures, only one way of perceiving is possible at any given instant.

Up until now we have considered the z coordinate of the surface to represent the most likely interpretation(s) given a (part of a) story, the x and y coordinates being some rough measure of the extent to which the two possible meanings are developed in the story. In jokes and humor that stimulate laughter (and generally *only* in this case), it may sometimes be more natural to take the z coordinate to be instead a rough measure of physiological excitation. We still have a quantity that depends on two factors, that has an abrupt jump (drop), and that satisfies the general conditions of Thom's theorem. The theorem thus still applies, and the qualitative shape of the surface generated is the same. Now, however, we can interpret the laugh accompanying the punch line of a joke as a release of emotional energy brought about by the catastrophic drop in physiological excitation. As I mentioned at the begin-

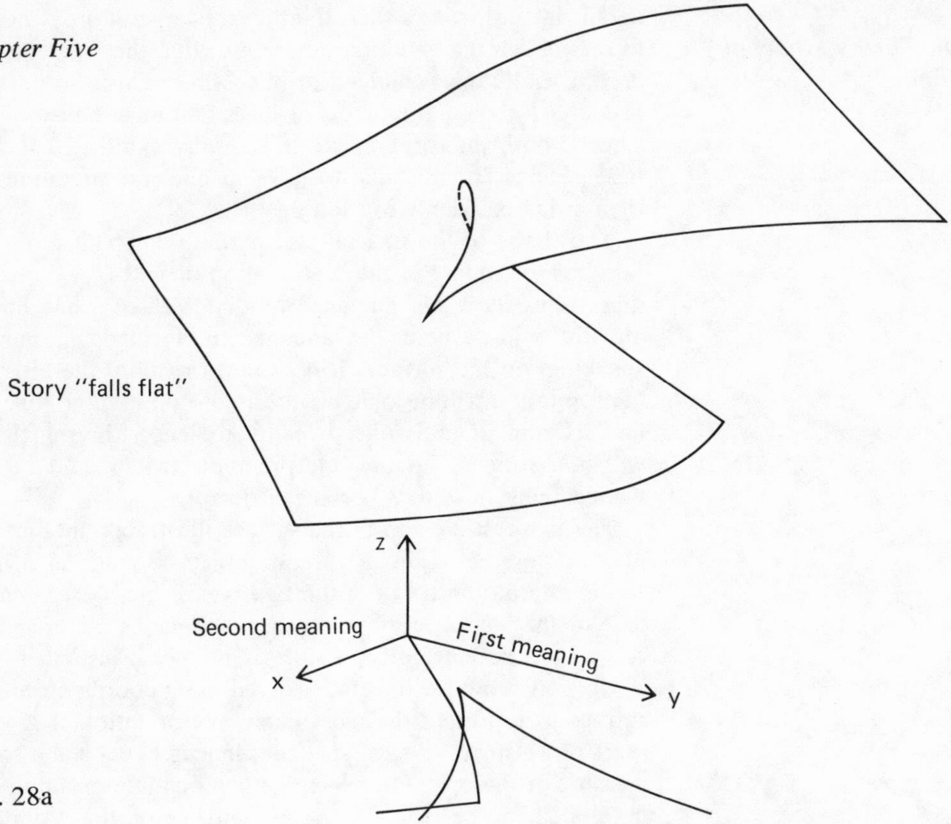

Fig. 28a

ning of this chapter, this emotional energy may result from overcoming a mild fear, releasing hostile feelings, expressing playfulness, or achieving satisfaction (fig. 30).

Thus the cusp catastrophe combines the cognitive incongruity theory and the various psychological theories of humor with the release theory of laughter—all in one parsimonious model. An incongruity or a pair of possible interpretations is of course necessary. This incongruity must, however, be such that its resolution releases emotional energy (from sexual

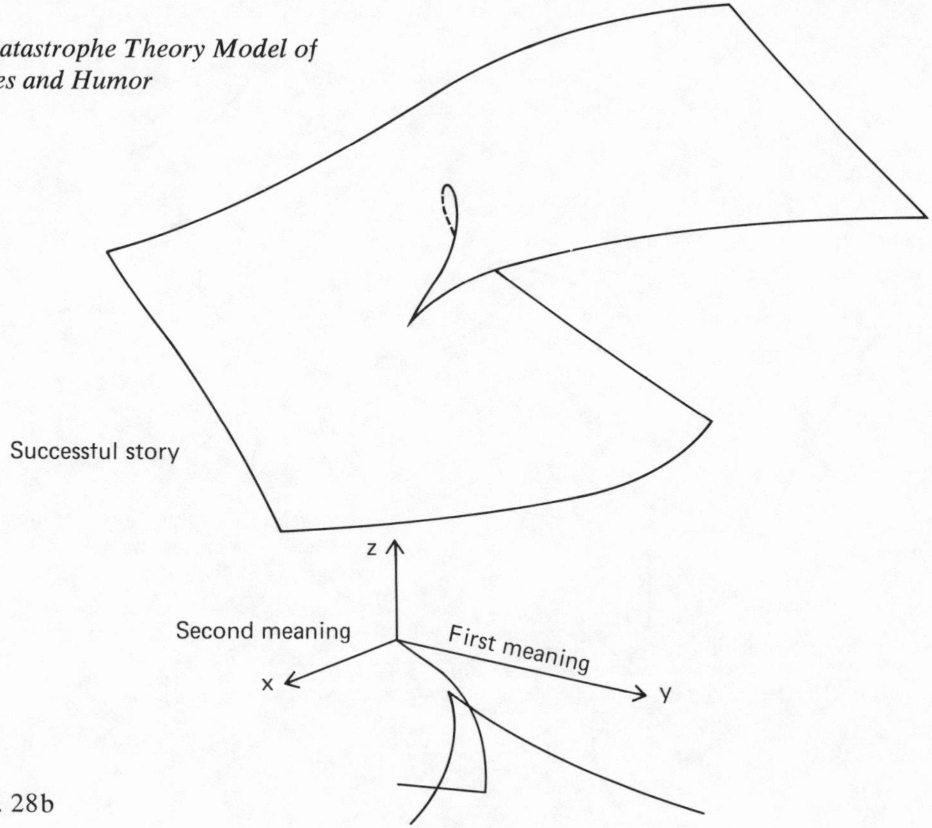

Fig. 28b

anxieties, "sudden glory," playfulness, or whatever). Moreover, the model is at least consistent with the derailment theory of humor, since the second (hidden) meaning (x coordinate) often depends critically on the context.

As I mentioned earlier, the model should be taken largely as a useful and suggestive mathematical metaphor for two reasons: accurately measuring the x, y, and z coordinates is usually very difficult and sometimes a matter of pure convention; and the model does not in general yield quantitative pre-

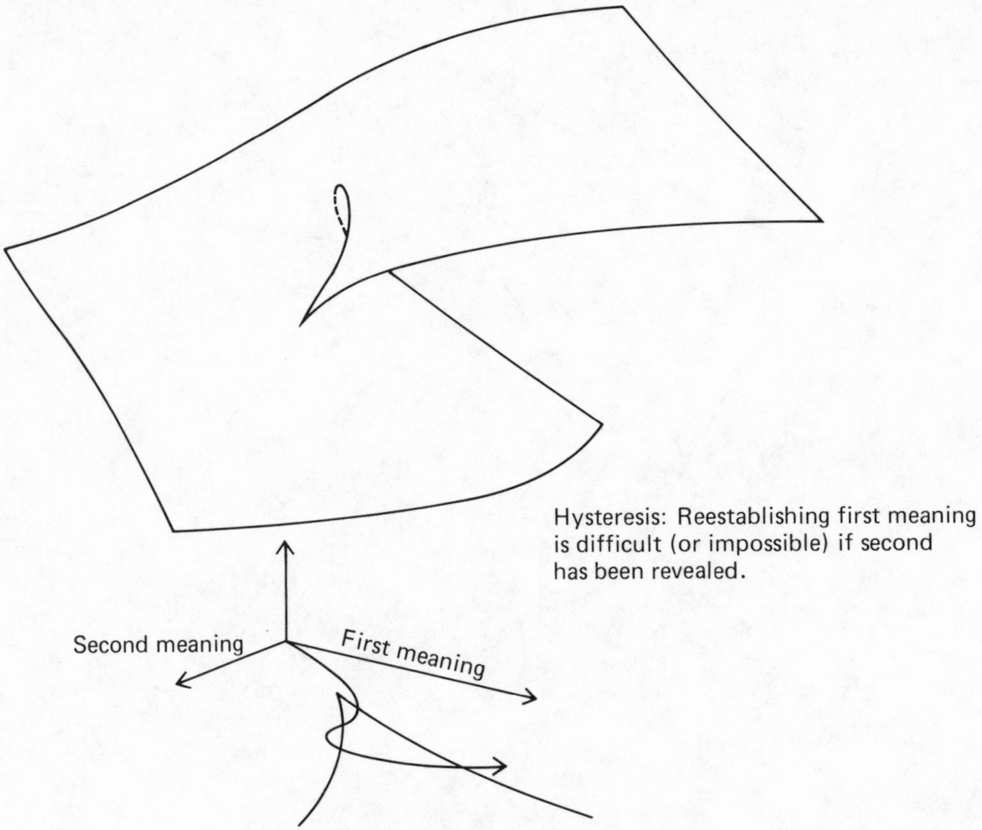
Hysteresis: Reestablishing first meaning
is difficult (or impossible) if second
has been revealed.
Second meaning
First meaning

Fig. 29

dictions but merely provides one with a qualitative shape. In certain restricted contexts these obstacles can, of course, be overcome, but generally they cannot. The model, in neatly combining the cognitive incongruity and the emotional climate aspects of humor with the release theory of laughter, provides one with at least the beginning of a pictorial insight into the structure of humor.

Although most simple jokes fit reasonably well into the model (with z as a measure of likely interpretation in the

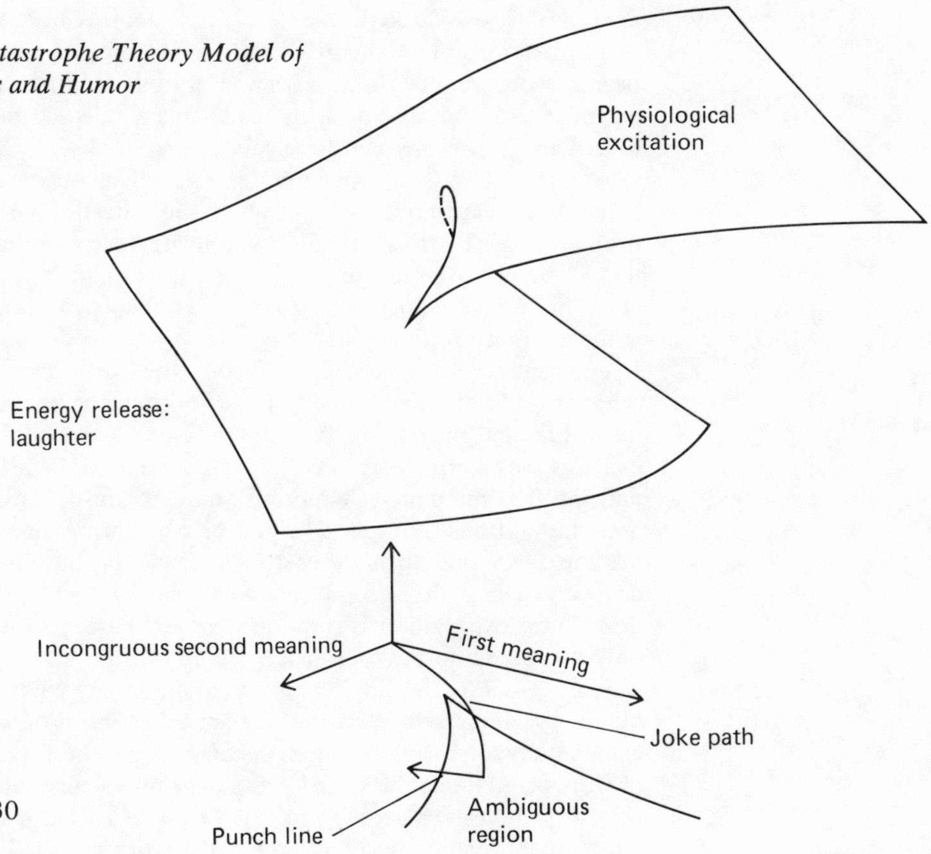

Fig. 30

case of jokes that do not provide laughter), there are some that must be bent a bit. Consider, for example, jokes that are all punch line—uncaptioned cartoons, caricatures, exaggerated gestures, unexpected pratfalls, sudden and surprising noises, even magic tricks. The model seems to break down, since there is apparently no development of one interpretation followed by a switch to a second interpretation. In these cases we can simply understand what is standard and conventional as a tacitly assumed interpretation and deviations from it (pratfalls, unexpected remarks, and so on) as the second "interpretation." The energy or tension released by laughter need not be assumed to have been built up but rather can be assumed to be always present, ready to be released at any time. This is perhaps more realistic in any case.

A few additional implications of the model ought to be mentioned. The model suggests that the catastrophic drop brought about by the punch line will be greater (more laughter) if there is a large gap between the upper and lower layers. This is likely to be the case in areas such as sex and authority that are surrounded by anxiety. Most old-time burlesque routines, for example, were concerned with one or the other or both. (More laughter, of course, does not necessarily mean funnier: nitrous oxide is not funny.) The model also explains why phrases such as "esthetically clumsy" and "big and fat" are marginally humorous while "clumsily esthetic" and "fat and big" are not. The first pair of phrases starts on the top layer and then drops to the lower layer, releasing energy, whereas the second pair goes in the opposite direction. Little expectation is developed in any case, so there is only a short jump between layers.

Finally, before developing a catastrophe theory analysis of humor involving self-referential metacues, let me propose an explanation for a widely noticed phenomenon, the funniness of words containing the sound *k* and their prevalence in (scatological) jokes. The reason for this, I think, is that *k* has a puncturing, debunking sound and is therefore especially appropriate in the punch line of a joke, where its onomatopoetic effect reinforces the catastrophe-producing punch line.

As an example (and to include at least one new joke in this chapter), consider the following story. A Greek regularly eats breakfast in a Chinese restaurant, where he always orders two fried eggs. The Chinese waiter always serves him politely, saying, "two flied eggs, sir." Finally, after years of this, the Greek gets fed up and explodes, "You idiot, learn to speak English. Two FRIED eggs, not two FLIED eggs. Understand?

Two FRIED eggs, two FRIED eggs!" The next morning the Chinese waiter serves him his eggs, saying, very politely, "two FRIED eggs . . . you Gleek plick."

In chapter 3 I showed how a version of the Cretan paradox is implicit in joke-telling, play, theater, and other endeavors in which there are metacues that belie the content of whatever is being said or done. This situation can be modeled by another much simpler catastrophe that results when a behavior (or any quantity) has a discontinuous jump, depends on just a single other factor, and satisfies certain mild general conditions. Given some numerical value for this other factor, say w, we will denote by $g(w)$ the most likely behavior z associated with it; $z = g(w)$. Thom's classification theorem states that the only possible graph for such a catastrophe must look qualitatively like the simple two-dimensional curve in figure 31 (two-dimensional, since z depends on just one factor and not on two as in the cusp catastrophe).

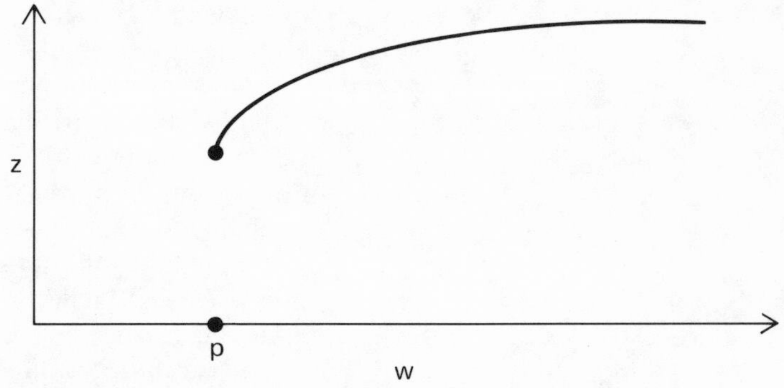

Fig. 31

If w is larger than p, then $z = g(w)$ is on the curve. As w decreases, however, a discontinuous jump occurs in the value of z at p as it drops precipitously. Similarly, if w is smaller than p, then $z = g(w)$ is low. But as w increases a discontinuity occurs in the value of z at the point p, where it jumps up to the curve.

We can apply this so-called fold catastrophe to the Cretan paradox in the following way. Take w to be a rough measure of the "realism" of the story. By this I mean a measure of the extent to which the story (or some initial part of it) is intended to be taken seriously at any given time. The story or joke comprises not only its content but also the manner in which it is presented. Thus, if during the story the storyteller winks or assumes a strange dialect, the value of w decreases. On the other hand, if the story is developed for a while with a certain internal consistency, w increases. We may take z, as in the case of the cusp catastrophe, to be some rough measure of physiological excitation. (Other interpretations of z are also possible and indeed more natural in more cerebral humor. The possible dependence of z on factors other than w will, for the time being, be ignored. I will return to both these matters shortly.) The equation $z = g(w)$ gives the most likely value of z corresponding to any given w. There is, of course, no unique way to assign values to z and w, but any reasonable convention will yield the same qualitative shape for these curves and surfaces.

Now, as discussed in chapter 3, any self-referential metacue induces an oscillation in the understanding of the story. If it is true, then it is false. If it is false, then it is true. As the story develops (w gradually increasing), the listener, attending to it, unconsciously becomes involved in it, and there is an abrupt jump in z. On the other hand, a small decrease in w (a slight inflection to the voice, say) may bring about a large drop in z as the listener momentarily consciously realizes that the story, joke, or play is "make-believe." By Thom's theorem, the only possible graph for such a discontinuity is that of the fold catastrophe (remember, in general z depends on factors other than w, but for now these are being ignored or being kept constant). Thus, as the story is being told (fig. 32) the listener oscillates between (1) following it, taking it seriously, and thereby becoming aroused, and (2) responding to the metacues, thereby realizing the story is make-believe, and becoming deflated. This phenomenon of mental oscillation partially accounts for the pleasant tension associated with good joke-telling, theater, and play.

As I mentioned in chapter 3, the humor of a joke results from the punch line together with this pleasant tension. Thus a more accurate model of the structure of a joke or humorous

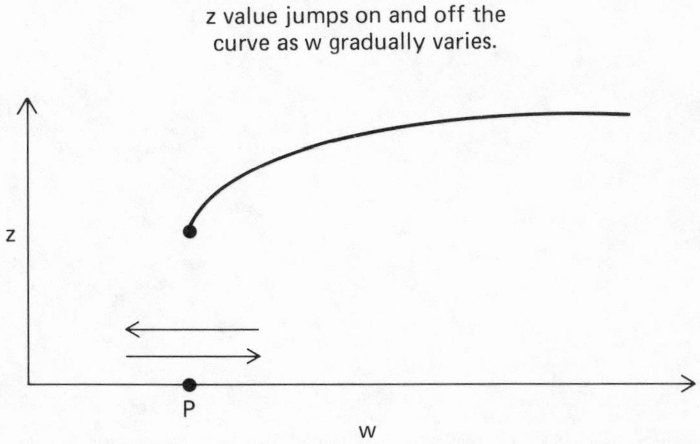
z value jumps on and off the
curve as w gradually varies.
z
P
w

Fig. 32

story would be one that somehow combined both these elements—some combination of the cusp and fold catastrophes. Here z would be a function of the three factors x, y, and w and would be the (a) most likely behavior associated with the triple (x, y, w). Thom's theorem here specifies that any quantity like z that has a discontinuity satisfying general conditions analogous to those in the two-factor case, and that depends on *three* other factors, *must* have a certain, quite distinctive graph known as the swallowtail catastrophe.

Since the graph is four-dimensional, it cannot easily be visualized. We can, however, draw a three-dimensional picture that is analogous to the cusp curve in the cusp catastrophe; that is, it indicates where the jump occurs (fig. 33). A joke path without the complication of the self-referential metacues

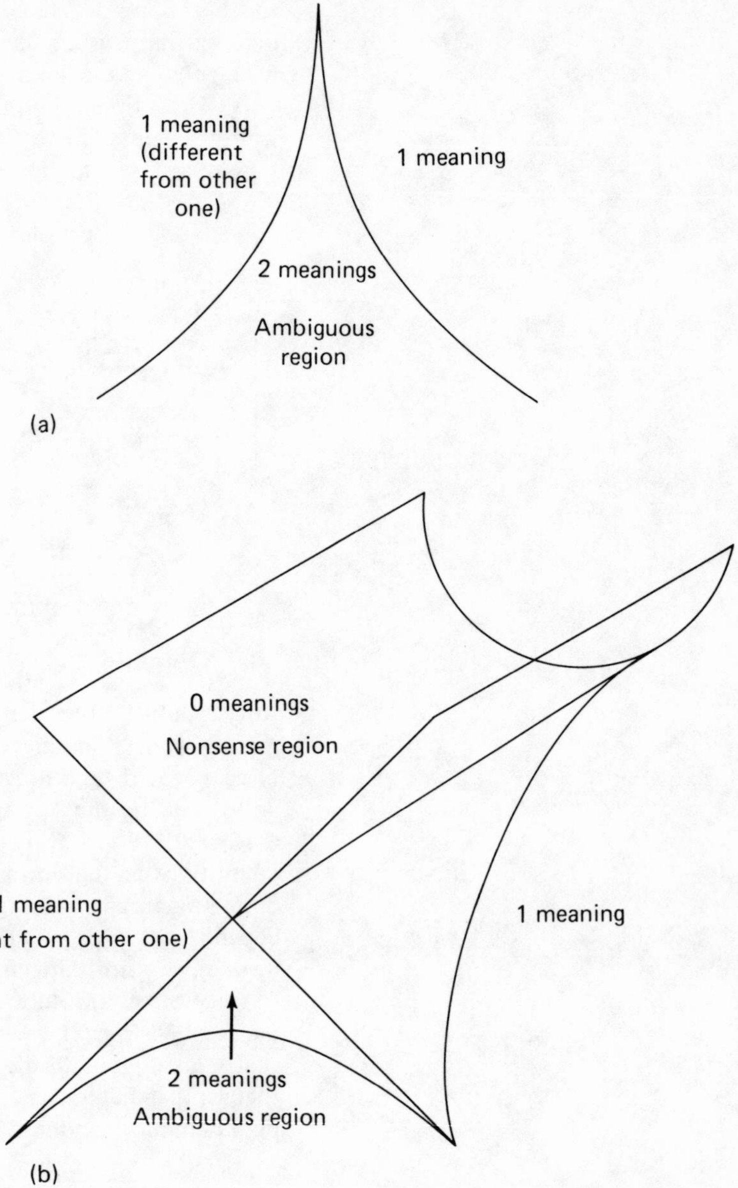
1 meaning
(different
from other
one)
1 meaning
2 meanings
Ambiguous
region
(a)
0 meanings
Nonsense region
1 meaning
(different from other one)
1 meaning
2 meanings
Ambiguous region
(b)

Fig. 33

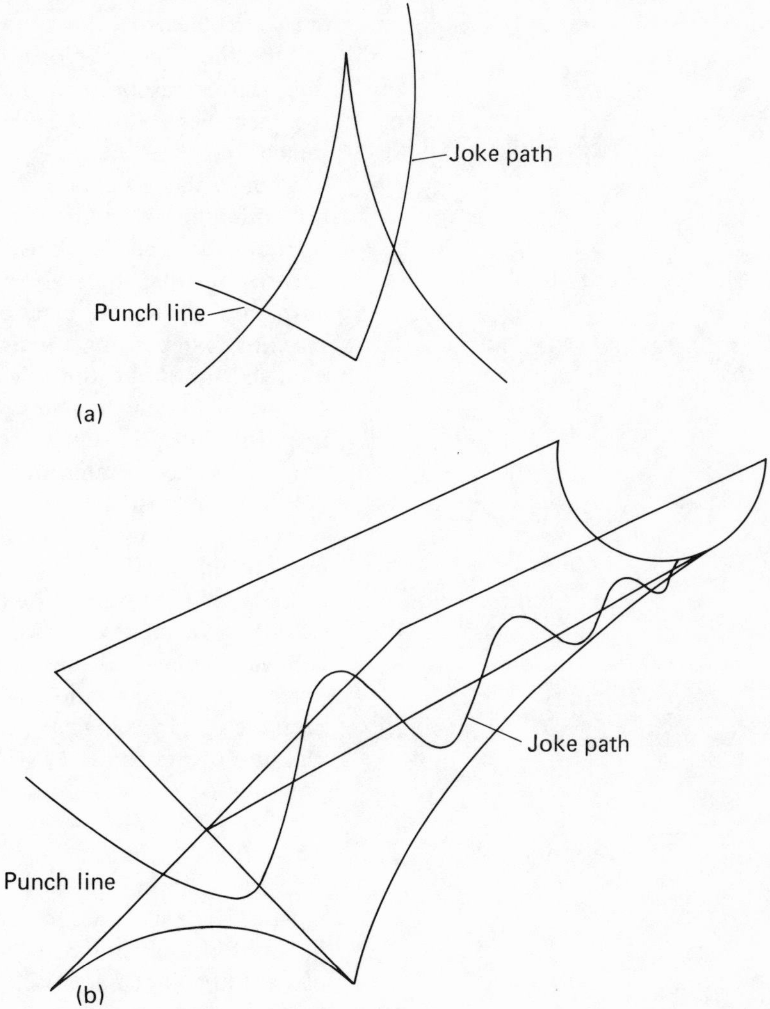

Fig. 34

is a line that starts at the origin, crosses the right side of the cusp into the ambiguous region, then cuts to the left at the punch line. The joke path in the case of the swallowtail catastrophe (fig. 34) begins at the origin, rises and falls a few times as indicated (alternating between the ambiguous region and the "nonsense" region), then cuts to the left for the punch line; this corresponds to small oscillations in tension due to the metacues followed by a larger release occurring at the punch line.

Thom's theorem, one should note, classifies *all* possible discontinuities in quantities satisfying the conditions mentioned above and dependent on not more than four factors. It turns out that there are only four in addition to the fold, cusp, and swallowtail catastrophes—seven in all. More complex jokes depending on four factors could conceivably be modeled by one of the other more intricate catastrophes. It is amusing to imagine paths on complicated convolutions with fiendishly intricate self-intersections as representing the skeletons of various humorous stories. Restriction to just one behavior dimension is not necessary, nor is our choice of physiological excitation as that behavior dimension the only one possible.

In fact, in the case of jokes, drama, and so forth involving qualifying metacues, it may be more natural to consider two behavior dimensions, one, z_1, concerning as before either the perceiver's interpretation of the ambiguous story or some measure of physiological excitation, and the other, z_2, some measure of the perceiver's judgment of the "realism" of the presentation (as opposed to the presenter's intentions regarding the "realism," w). Again, I am getting metaphorical, but what of it? Much of our understanding is metaphorical rather than scientific in a narrow sense. But, clearly, much work must be done if we are to achieve anything beyond metaphor.

Several literary analyses of comedy are vaguely reminiscent of the foregoing analysis of jokes in terms of the cusp catastrophe. The critic Northrop Frye, for example, writes that in a classical comedy "what normally happens is that a young man wants a young woman, that his desire is resisted by some opposition, usually paternal, and that near the end of the play some twist in the plot enables the hero to have his will. . . .

At the end of the play the device in the plot that brings hero and heroine together causes a new society to crystallize around the hero, and the moment when this crystallization occurs is the point of recognition in the action, the comic discovery . . . the obstacles to the hero's desire, then, form the action of the comedy, and the overcoming of them the comic resolution."

The comic discovery or resolution corresponds to the play's punch line, so to speak. In a very rough sense, then, we might characterize the structure of comedy as a very large cusp catastrophe (actually a swallowtail catastrophe), corresponding to the structure Frye describes and including within it a collection of smaller catastrophes, the jokes or laugh lines of the comedy. (We thus have provided a second sense in which a comedy can be referred to as a catastrophe.) Obviously this analysis is simplistic, but it is nevertheless suggestive. A comedy, even a modern one, seems to require some sort of obstacle-struggle-happy-resolution pattern without which, no matter how funny it may be, one is a little reluctant to call it a comedy. The jokes must be integrated into a coherent comedic structure, not just enumerated.

Finally, note that the placement of this chapter on catastrophe theory at the climax, so to speak, of this book makes its structure a sort of (self-referential) cusp catastrophe. This chapter on catastrophe theory is the punch line of the larger cusp catastrophe that is the book.

6

Odds and the End

In this last chapter I will place humor (specifically the logic of humor) in a broader perspective and briefly indicate some extensions and ramifications of what I have said about it.

To these ends, let us first consider cognitive psychology, that branch of psychology that deals with cognition and the intellectual process. It has only relatively recently separated itself from the umbrella discipline of natural philosophy, and it has as its main concern issues (such as perception, memory, concept formation, problem-solving, and consciousness) that have traditionally been known as "philosophical problems." Even today much conceptual analysis relevant to cognitive psychology is done by philosophers.

Recently, however, important advances have been made in the development of cognitive psychology by psychologists and others who have shaken off the behaviorist tendency to reduce all mental phenomena to stimulus-response sequences of some form. For example, Noam Chomsky, whom I have already discussed briefly in chapter 4, has posited a quasi-innate mental "mechanism" to account for the fact that all human beings learn to speak on the basis of a very meager collection of data and for the similarities among human languages of the transformations that change the deep structure of a statement in a given language into an acceptable surface structure. We must hypothesize some internalization of these transformational rules to account for why people can formulate (and understand) original statements that have never before been spoken. (When and how these rules are used are not, however, subject to rules.)

In a somewhat different direction, the Swiss psychologist Jean Piaget, whom I also mentioned earlier, has investigated the cognitive development of children and in particular the maturation of such basic skills as correspondence, subordination, seriation, classification, conservation, and symmetry recognition. These skills mature at different stages in a child's development and can therefore be used to measure that development. A child's development of geometric notions is an example of Piaget's findings. Topological properties are mastered first (betweenness, connectedness), then projective properties are learned (triangularity, equivalence of circles and ellipses), and finally comes an understanding of metric properties (length, angles). This order, though logically superior, is opposite to the historical development of geometry. Perhaps topology should be taught in primary school. In addition to the theories of Chomsky and Piaget, there has of course been much important work on concept formation, short- and long-term memory, mental organization of information, problem-solving techniques, and so on.

I have introduced these examples of advances in cognitive psychology and the earlier expositions of certain philosophical matters to provide some perspective on the subject of this book; the *logic* of humor seems to contribute substantially to the broader subject of cognitive psychology as well as indirectly to philosophical analyses of human language and action. Since humor is such a complex and human phenomenon, any understanding of it will necessarily enrich our understanding of thought in general. Moreover, since laughter often accompanies jokes and humor, there is here an obvious behavioral manifestation of thought.

The logical and linguistic devices I have discussed in chapters 2, 3, and 4, and that I have employed to bring to mind two interpretations of some phenomena, should be useful more generally in cognitive psychology and perhaps even in research on artifical intelligence. Certain ideas from model theory, for example, might be fruitful in understanding the semantics of (parts of) natural languages (see Fodor 1975; Katz 1971; Montague 1974). Similarly, the notion of statement levels and convoluted hierarchies is proving useful not only in humor but more generally. Its relation to the semantics of so-called nested clauses in transformational grammar, for example, is very interesting.

More general applications of self-referential notions are also plentiful. As I mentioned earlier, the schizophrenogenic effect of versions of the Cretan or liar paradox has been studied by Bateson, Laing, and others. If the way a statement (or more generally a set of statements) is made belies its content, a paradox results—a paradox that can have very deleterious behavioral consequences if it affects concerns of great importance to a person. The importance of metacues in setting up a frame in humor, and in art in general, has also been mentioned, yet it seems that there is still much work to be done on both the behavioral and the artistic consequences of self-reference.

Catastrophe theory has been used by Zeeman (1972) to model brain processes and (sketchily at least) by Thom (1975) to model speech. I have employed it in this book to model the structure of humor, but applications to other cognitive processes are possible. The property of divergence was used in the aggression example of chapter 5 to explain why two slightly different paths to the same fear-rage coordinates might produce very different behaviors. It could also be used to account for the well-known fact that, when asked to choose the word that does not belong in the sequence {skyscraper, cathedral, temple, prayer} people usually pick *prayer*, whereas in the sequence {prayer, temple, cathedral, skyscraper} they usually pick *skyscraper*. There are doubtless other such opportunities for application of catastrophe theory to problems in cognition.

There remain, of course, many open questions to ponder. Is there (or can there be) a typology of joke structures? What connection, if any, do these joke structures have to other linguistic structures? What is the relation between consciousness and the self-referential paradoxes (and nonparadoxes)? Can the catastrophe theory model be integrated with computer models of thought? Will your sense of humor survive this book?

Leaving these questions and cognitive psychology, I would like to discuss the relativity of the notion of incongruity. My "formula" for humor has been "a perceived incongruity with a point, in an appropriate emotional climate." (Incongruity, as I stated in chapter 1, is intended in a wide sense, comprising the following oppositions: expectation versus surprise, the mechanical versus the spiritual, superiority versus incompetence, balance versus exaggeration, and propriety versus vulgarity.) A problem with this formulation is that *any* person, object, or situation is incongruous in *some* sense or with respect to *some* standard of appropriateness. Even a green ball lying on a table could, given a certain context, be considered incongruous. Perhaps it is 1 January 2001 and the ball has, to some foreign visitor, just changed colors from grue to bleen.

Thus notions like "congruous," "appropriate," and "relevant," which I have used throughout, are almost always relative to some cultural milieu and language and dependent on a particular context. So also is the notion of the "point" of a joke. Different cultures, subcultures, and individuals in varying contexts consider different actions, situations, combinations of attributes, and so forth, to be incongruous. From these trivial observations follows the equally trivial but nevertheless interesting fact that much of what is humorous varies, at least in degree, from culture to culture and from context to context. "In" jokes of various kinds, jokes depending on the peculiarities of a language, topical and political humor, and so forth, are obvious examples. (I am speaking here of the relativity of the content of humor, *not* of its structure, which I am assuming to be universal.)

Sometimes a clash *between* subcultures produces humor. The frame of reference of a subculture is skewed or out of kilter with respect to that of the "dominant" (sub)culture. If the cultures are not too different, this skewness enables members of the minority group to view the dominant culture from a certain (slightly jaundiced) perspective from which social incongruities are more glaring. A disproportionate share of American comedians, for example, are Jewish or black; a disproportionate share of British comedians are Irish. To be a comedian, one must be sensitive to one's social environment yet sufficiently alienated so as to see (parts of) it from the outside—from the metalevel, if you will. This partial alienation allows a more abstract view of things, usually a precondition for employing the formal devices considered in this book. Being too far removed, however, is inconsistent with empathy, with being sensitive to the proper "emotional climate." It is probably because they must both empathize with and understand the values of their society as well as be able to step beyond them that comedians are often perceived either as very warm and sympathetic or as aloof and scornful. They are probably both—more human (but not necessarily more humane) than most people.

Of course some incongruities are less culture-bound than others, and there does seem to be an almost-universal class of such incongruities. By this I mean that some ways of ordering the world are so basic—perhaps the elementary laws of logic and arithmetic or perhaps Piaget's basic cognitive skills of conservation, seriation, and so on—that to violate them is incongruous in any culture. Many of the jokes considered in this book are based on this. (Wife to husband: Should I cut your meat loaf into four or eight pieces? Husband: Four;

I'm trying to lose weight. Or, Old man to second old man: I like taking long walks by myself. Second old man: Me too. Let's go.)

The last point I would like to make concerns the unlikely topic of scientific development. Thomas Kuhn in his *Structure of Scientific Revolutions* (1970) has argued that different scientific theories (say Ptolemy's astronomy versus Copernicus's, or Newton's theory of gravitation versus Einstein's) do not always develop in a gradual and cumulative manner. More precisely, he claims that a scientific theory (in any field, not just physics) *normally* develops in a more or less cumulative way, new results gradually being added and old ones being slightly modified. Sometimes, however, a theory slowly becomes inadequate (cf. Ptolemy or Newton), observations are made that are anomalous and incongruous, and many explanations become unconvincing and ad hoc. Often after a time a new theory suddenly appears (cf. Copernicus and Einstein) that is incommensurable with the earlier one. New ideas arise, old terms take on radically new meanings, previously unnoticed relations become important, and so forth. This relatively sudden development of the new theory (or paradigm) Kuhn calls a "scientific revolution."

It is fitting (scientific? funny?) to note that, given certain plausible factual assumptions and using Thom's theorem, we can conclude that the structure of a scientific revolution is similar to the structure of a joke. It is a swallowtail catastrophe, with x and y being the extent of observational support for the alternative theories, w being a measure of "meaningful raw observations," and $z = f(x, y, w)$ being the interpretation(s) most likely accepted given x, y, and w. The path of a scientific revolution is similar to the path of a joke in a swallowtail catastrophe. The anomalous (incongruous) observations leading to the "scientific revolution" correspond to the joke's punch line as described in the last chapter.

Recall from chapter 1 Arthur Koestler's claim in *The Act of Creation* (1964) that the logic of the creative process is the same in art, science, and humor, and that only the "emotional climate" differs. This last observation regarding scientific revolutions and jokes is both further evidence for and a refinement of Koestler's vague but fertile and suggestive thesis. Much of the present book in fact is, as I wrote in chapter 1, a development of this thesis in the case of mathematics, considered as an art, and of humor, especially cognitive humor.

Finally, though there are better visions of the world than that of a huge self-referential joke containing within it countless smaller jokes, something like that vision is at the source of this book. It has been my pleasure to increase by one the number of jokes in the world.

References

Allen, Woody. 1972. *Getting even*. New York: Warner Books.

Barker, Stephen. 1964. *Philosophy of mathematics*. Englewood Cliffs, N.J.: Prentice-Hall.

Barrick, M. E. 1974. The newspaper riddle joke. *Journal of American Folklore* 87:253–57.

Bateson, Gregory. 1958. The message "this is play." In *Group processes: Transactions of the second conference*, ed. B. Schaffner. New York: Josiah Macy, Jr., Foundation.

Baudelaire, Charles Pierre. 1962. *Curiosités esthetiques*. Paris: Garnier.

Beattie, James. 1776. An essay on laughter, and ludicrous composition. In *Essays*. Edinburgh: William Creech.

Bellman, Richard. 1969. Humor and paradox. Technical report, National Institutes of Health.

Bergson, Henri. 1911. *Laughter: An essay on the meaning of the comic*. New York: Macmillan.

Carroll, Lewis. 1946. *Alice's Adventures in Wonderland* and *Through the Looking Glass*. New York: Grosset and Dunlap.

Chapman, A. J., and Foot, H. C., eds. 1976. *Humor and laughter: Theory, research, and applications*. London: Wiley.

———, eds. 1978. *It's a funny thing, humor*. Oxford: Pergamon.

Chomsky, Noam. 1968. *Language and mind*. New York: Harcourt, Brace, and World.

Dupréel, P. 1928. The sociology of laughter. *Revue Philosophique* 106:213–60.

Eastman, Max. 1936. *Enjoyment of laughter*. New York: Simon and Schuster.

Fodor, Jerry. 1975. *The language of thought*. New York: Crowell.

Freud, Sigmund. [1905] 1960. *Jokes and their relation to the unconscious*. New York: Norton.

Fry, W. F., Jr. 1963. *Sweet madness: A study of humor*. Palo Alto, Calif.: Pacific Press.

Frye, Northrop. 1958. The structure of comedy. In *Eight great comedies*, ed. S. Barnett. New York: New American Library.

Goldstein, J. H., and McGhee, P., eds. 1972. *The psychology of humor*. New York: Academic Press.

Goodman, Nelson. 1965. *Fact, fiction, and forecast*. New York: Bobbs-Merrill.

Halmos, Paul. 1960. *Naive set theory*. New York: Van Nostrand.

Hazlitt, William. 1819. On wit and humor. In *Lectures on the English comic writers*. London: Taylor and Hessey.

Hempel, Carl. 1965. *Aspects of scientific explanation*. New York: Free Press.

Hobbes, Thomas [1651] 1914. *Leviathan*. London: Dent.

Kant, Immanuel. 1892. *Kant's kritik of judgment*. Translated by J. H. Bernard. London: Macmillan.

Katz, Jerrold J. 1971. *The underlying reality of language*. New York: Harper Torchbooks.

Kliban, B. 1976. *Never eat anything bigger than your head*. New York: Workman.

Kneebone, G. T. 1963. *Mathematical logic and the foundations of mathematics*. London: Van Nostrand.

Koestler, Arthur. 1964. *The act of creation*. London: Hutchinson.

Kripke, Saul. 1975. Outline of a theory of truth. *Journal of Philosophy*, December, pp. 690–716.

Kuhn, Thomas. 1970. *The structure of scientific revolutions*. 2d ed. Chicago: University of Chicago Press.

La Fave, Lawrence. 1972. Humor judgments as a function of reference group and identification classes. In *The psychology of humor*, ed. J. H. Goldstein and P. McGhee. New York: Academic Press.

———. 1978. Ethnic humor: From paradoxes towards principles. In *Humor and laughter: Theory, research, and applications*, ed. A. J. Chapman and M. C. Foot. London: Wiley.

Laing, R. D. 1970. *Knots*. New York: Vintage Books.

Lyndon, Roger C. 1966. *Notes on logic*. New York: Van Nostrand.

Malcolm, N. 1958. *Ludwig Wittgenstein: A memoir*. London: Oxford University Press.

McGhee, P. 1978. A model of the origins and early development of incongruity-based humor. In *Humor and laughter: Theory, research, and applications*, ed. A. J. Chapman and M. C. Foot. London: Wiley.

Meredith, George. 1918. *An essay on comedy*. New York: Charles Scribner's Sons.

Milner, G. B. 1972. Homo ridens: Towards a semiotic theory of humor and laughter. *Semiotica* 1:1–30.

Moise, E. E. 1963. *Elementary geometry from an advanced standpoint*. Reading, Mass.: Addison-Wesley.

Monro, D. H. 1951. *Argument of laughter*. Melbourne: Melbourne University Press.

Montague, Richard. 1974. *Formal philosophy*, edited by Richmond H. Thomason. New Haven: Yale University Press.

Paulos, John A. 1978*a*. Applications of catastrophe theory to semantics. *Notices of the American Mathematical Society* 25 (January):A-173.

———. 1978*b*. The logic of humor and the humor in logic. In *Humor and laughter: Theory, research, and applications*, ed. A. J. Chapman and M. C. Foot. London: Wiley.

———. 1979. A model-theoretic account of confirmation. *Notre Dame Journal of Formal Logic*. 20:451–58.

Piaget, Jean. 1952. *The origins of intelligence in children*. New York: International Universities Press.

Piddington, Ralph. 1933. *The psychology of laughter: A study of social adaptation*. London: Figurehead; reissued, New York: Gamut Press, 1963.

Pitcher, George. 1966. Wittgenstein, nonsense, and Lewis Carroll. *Massachusetts Review*, August, pp. 591–611.

Quine, W. V. O. 1968. Paradox. *Scientific American*, April, pp. 84–95.

Rogers, H. 1967. *Theory of recursive functions.* New York: McGraw-Hill.

Rosten, Leo. 1968. *The joys of Yiddish.* New York: McGraw-Hill.

Russell, B., and Whitehead, A. N. 1910. *Principia mathematica.* Cambridge.

Saussure, F. 1931. *Cours de linguistic générale.* Paris: Payot.

Schopenhauer, Arthur. 1883. *The world as will and idea.* London: Trubner.

Shoenfield, J. R. 1967. *Mathematical logic.* Reading, Mass.: Addison-Wesley.

Shultz, T. R. 1976. A cognitive-developmental analysis of humor. In *Humor and laughter: Theory, research, and applications*, ed. A. J. Chapman and M. C. Foot. London: Wiley.

Suls, J. 1972. A two-stage model for the appreciation of jokes and cartoons. In *The psychology of humor*, ed. J. H. Goldstein and P. McGhee. New York: Academic Press.

Tarski, Alfred. 1936. Der Wahrheitsbegriff in formalisierten Sprachen. *Studia Philosophica* 1:261–405.

Thom, René. 1975. *Structural stability and morphogenesis.* Reading, Mass.: W. A. Benjamin.

Turing, A. M. 1950. Computing machinery and intelligence. *Mind* 59:433–60.

Warnock, G. J. 1966. *English philosophy since 1900.* New York: Oxford University Press.

Wittgenstein, Ludwig. 1953. *The philosophical investigations.* Oxford: Blackwell.

———. 1958. *Remarks on the foundations of mathematics.* New York: Macmillan.

Zeeman, E. C. 1972. Catastrophe theory in brain modelling. *Conference on neural networks.* Trieste: ICTP.

———. 1976. Catastrophe theory. *Scientific American*, April, pp. 65–83.

Index

Iteration: in humor, 35–38; in mathematics, 32–35